SPACE SCIENCE, EXPLORATIO

NASA's MANAGEMENT OF COMMERCIAL CREW AND CARGO OPERATIONS

ASSESSMENTS

EILEEN PURCELL
EDITOR

New York

NOTICE TO THE READER

Library of Congress Cataloging-in-Publication Data

ISBN: 978-1-63484-065-1

Published by Nova Science Publishers, Inc. † New York

NASA's MANAGEMENT OF COMMERCIAL CREW AND CARGO OPERATIONS

ASSESSMENTS

SPACE SCIENCE, EXPLORATION AND POLICIES

Additional books in this series can be found on Nova's website under the Series tab.

Additional e-books in this series can be found on Nova's website under the e-book tab.

CONTENTS

PREFACE

Since the retirement of the Space Shuttle Program in July 2011, the United States has lacked a domestic capability to transport crew and – until recently – cargo to and from the International Space Station (ISS or Station). Consequently, NASA has been relying on the Russian Federal Space Agency (Roscosmos) for crew transportation. In anticipation of the Shuttle's retirement, Congress and the President directed NASA to foster the commercial spaceflight industry as a means of developing domestic cargo and crew transportation capabilities to the Station. In November 2005, NASA created the Commercial Crew and Cargo Program Office and in 2011, activated a separate Commercial Crew Program Office to reflect the increased funding and priority for commercial crew. In June 2013, the Office of Inspector General (OIG) issued a report examining NASA's efforts to foster a commercial market for cargo resupply missions to the ISS. The report discusses NASA's funding over the past 7 years of SpaceX and Orbital to further development of spaceflight capabilities and, on a separate track, the Agency's contracts with the companies for a combined 20 cargo resupply missions to the ISS. As a complement to that report, this book examines NASA's efforts to pursue commercial crew capabilities.

In: NASA's Management of Commercial Crew ...   ISBN: 978-1-63484-065-1
Editor: Eileen Purcell   © 2016 Nova Science Publishers, Inc.

Chapter 1

NASA'S MANAGEMENT OF THE COMMERCIAL CREW PROGRAM[*]

NASA Office of Inspector General

ACRONYMS

ASAP	Aerospace Safety Advisory Panel
CCDev1	Commercial Crew Development Round 1
CCDev2	Commercial Crew Development Round 2
CCiCap	Commercial Crew Integrated Capability
CCtCap	Commercial Crew Transportation Capability
CPC	Certification Products Contact
FAA	Federal Aviation Administration
FAR	Federal Acquisition Regulation
FY	Fiscal Year
GAO	Government Accountability Office
ISS	International Space Station
NPR	NASA Procedural Requirements
OIG	Office of Inspector General
OMB	Office of Management and Budget

---

[*] This is an edited, reformatted and augmented version of a report, IG-14-001, issued by the National Aeronautics and Space Administration's Office of Inspector General, November 13, 2013.

OVERVIEW

The Issue

Since the retirement of the Space Shuttle Program in July 2011, the United States has lacked a domestic capability to transport crew and – until recently – cargo to and from the International Space Station (ISS or Station). Consequently, NASA has been relying on the Russian Federal Space Agency (Roscosmos) for crew transportation. Between 2012 and 2017, NASA will pay Roscosmos $1.7 billion to ferry 30 NASA astronauts and international partners to and from the ISS at prices ranging from $47 million to more than $70 million each. After 2017, NASA hopes to obtain transportation to the ISS from American companies.

In anticipation of the Shuttle's retirement, Congress and the President directed NASA to foster the commercial spaceflight industry as a means of developing domestic cargo and crew transportation capabilities to the Station. In November 2005, NASA created the Commercial Crew and Cargo Program Office and in 2011, activated a separate Commercial Crew Program Office to reflect the increased funding and priority for commercial crew. As of August 31, 2013, NASA has spent $1.1 billion on its commercial crew development efforts. NASA is currently working with three companies – The Boeing Company (Boeing), Space Exploration Technologies Corporation (SpaceX), and Sierra Nevada Corporation (Sierra Nevada) – using a combination of funded Space Act Agreements and contracts based on the Federal Acquisition Regulation (FAR) to develop commercial crew transportation capabilities.[1]

In June 2013, the Office of Inspector General (OIG) issued a report examining NASA's efforts to foster a commercial market for cargo resupply missions to the ISS.[2] In the report, we discussed NASA's funding over the past 7 years of SpaceX and Orbital to further development of spaceflight capabilities and, on a separate track, the Agency's contracts with the companies for a combined 20 cargo resupply missions to the ISS. As a complement to that report, this review examines NASA's efforts to pursue commercial crew capabilities.

NASA's Commercial Crew Program is currently at a critical stage of development, as Boeing, SpaceX, and Sierra Nevada are expected to complete their spacecraft designs within the next year. While the partners are responsible for developing the vehicles, they rely heavily on NASA funding to meet current schedule requirements. At the same time, NASA maintains responsibility for ensuring that the partners' launch systems and spacecraft meet Agency safety and operational requirements. All three partners achieved what amounts to a Preliminary Design Review prior to NASA's award of the latest round of Space Act Agreements in 2012 and have set an optimistic schedule for achieving what each company defines as a Critical Design Review of their systems by mid-2014.[3]

After completion of the initial two rounds of development using Space Act Agreements, NASA originally planned a two-phased, FAR-based acquisition approach to develop commercial crew capabilities. Phase 1 was to consist of firm-fixed-price contracts to multiple companies for integrated design and early development followed by a second round of firm-fixed-price contracts for additional development, testing, evaluation, and certification of the contractors' crew transportation systems. Thereafter, NASA planned to enter into individual FAR-based contracts for each crewed mission with the hope of beginning flights to the ISS in 2016.

For fiscal year (FY) 2012, NASA received $397 million for its Commercial Crew Program; less than half its $850 million request. In light of this development, in August 2012, NASA revised its Commercial Crew Acquisition Strategy to rely on Space Act Agreements rather than FAR-based contracts for the integrated design phase of the program. The Agency also delayed the expected completion date of the commercial crew development phase from 2016 to 2017.

Using funded Space Act Agreements rather than FAR-based contracts to develop new transportation capabilities has several potential benefits. First, because the partners share development costs and the agreements involve fewer regulations and require less oversight by NASA, they may reduce the Agency's cost of acquiring these capabilities. Second, because NASA does not impose specific requirements on the companies as part of the agreements, the commercial partners are free to develop spacecraft designs that will support the needs of both NASA and other potential customers. Third, NASA officials said they believe the greater flexibility offered by Space Act Agreements promotes creativity and innovation.

However, NASA's use of Space Act Agreements also poses risks, most prominently limiting its ability to dictate specific design and safety requirements during the development process. In addition, oversight of partners in a Space Act Agreement relationship is more challenging. Taken together, these constraints make it harder to ensure that the companies will ultimately produce spaceflight systems that meet Agency requirements and that NASA can be confident will safely carry its astronauts to and from the ISS.

Given the significance of the Commercial Crew Program to NASA's human spaceflight efforts, we assessed: (1) the progress of each of the commercial partners toward developing a certified crew capability and (2) the major challenges that must be addressed to successfully implement the Program. Details of the audit's scope and methodology are in Appendix A.

Results

Although NASA's commercial partners are making steady progress in their initial development, the Commercial Crew Program faces several obstacles that may hinder NASA's ability to transport astronauts to the ISS in commercially supplied vehicles by 2017. These include an unstable funding stream, challenges in aligning cost estimates with the Program schedule, challenges in providing timely requirement and certification guidance, and coordination issues with other Federal agencies. Failure to address each of these challenges in a timely manner could significantly delay the availability of commercial crew transportation services and extend U.S. reliance on the Russians for crew transportation to the ISS.

NASA's Commercial Partners Meeting Early Development Milestones, but Face Significant Future Challenges. While their development approaches vary, all three commercial crew partners are successfully executing their spaceflight development efforts in accordance with their Space Act Agreements and FAR-based contracts. Each partner has completed its preliminary space system design and is in the process of finalizing that design and conducting component testing. In December 2012, NASA awarded all three partners a Phase 1 FAR-based certification contract pursuant to which the partners are working to ensure their space systems meet NASA requirements. However, Critical Design Review and other key milestones historically associated with cost and schedule delays remain to be completed.

NASA Must Overcome Major Hurdles to Provide Commercial Crew Transportation on Current Schedule. We identified four significant challenges to NASA's spaceflight development program for crewed flights to the ISS: unstable funding, alignment of cost estimates with the Program schedule, challenges in providing timely requirement and certification guidance, and spaceflight coordination issues with other Federal agencies.

Funding Instability Jeopardizes NASA's Plans to Begin Certified Crew Missions to the ISS by 2017. The Program received only 38 percent of its originally requested funding for FYs 2011 through 2013, bringing the current aggregate budget shortfall to $1.1 billion when comparing funding requested to funding received. As a result, NASA has delayed the first crewed mission to the ISS from FY 2015 to at least FY 2017. For FY 2013, Commercial Crew Program managers had already expected less than the requested $830 million and based their planning on a funding level of $525 million. The combination of a future flat-funding profile and lower-than-expected levels of funding over the past 3 years may delay the first crewed launch beyond 2017 and closer to 2020, the current expected end of the operational life of the ISS.[4]

Aligning Cost Estimates with Program Schedule May Enhance Management and Improve Transparency of Commercial Crew Program. Although NASA's Commercial Crew partners have completed their preliminary spacecraft designs, NASA managers have yet to develop a life cycle cost estimate for the Program. Life cycle costs are the anticipated annual costs of a program throughout its life, from preliminary design through the end of operations. According to Agency guidance, program managers should develop a life cycle cost estimate before Preliminary Design Review has been completed to help avoid future development delays caused by a lack of funding.[5] However, Program officials told us that this guidance does not apply to programs developed using Space Act Agreements and that developing such estimates for these type of programs is difficult due to a number of factors, including lack of historical cost data. Accordingly, rather than life cycle cost estimates, NASA developed independent government cost estimates projecting the overall cost to complete development and certification of the complete system based upon data supplied by the commercial partners, NASA insight into each company's program, and historical data from other NASA and Air Force rocket programs. Although these estimates provided some indication of the cost to develop a viable commercial crew transportation capability, they do not provide the same level of fidelity regarding overall costs as a life cycle cost estimate.

We acknowledge the difficulties in developing a comprehensive life cycle cost estimate for a program using a unique and still evolving acquisition approach for which there is limited comparable historical data. However, without a detailed cost estimate for each year of the program based upon a complete analysis of each subcomponent of the program over time, it is difficult for NASA to calculate how much funding is required each year given that costs over time can fluctuate significantly. Moreover, improved guidance requiring detailed cost estimates earlier in program development may help enhance the transparency of NASA's congressional budget submissions. To this point, members of Congress and the NASA Aerospace Safety Advisory Panel have stated that the lack of this type of information makes it difficult to know with any level of confidence the amount of government money required to develop a viable commercial crew transportation capability.[6] In our judgment, because NASA now has some historical cost data from the Commercial Crew Program, it is better positioned to develop improved guidance for cost estimating when systems are developed using Space Act Agreements.

Timeliness of NASA's Process for Considering Alternative Human Rating Requirements Remains a Concern. NASA has been challenged to provide partners with timely requirement and certification guidance during the current period of development. While NASA used Space Act Agreements to encourage innovative designs by its partners during system development, it is using FAR-based firm-fixed-price contracts for the certification process. As such, the partners must rely on timely guidance and information from NASA, and delays could affect their ability to move forward with production. As a result, the certification contracts include a 90-day goal for NASA to respond to partner requests for requirement and certification guidance. However, the general Program Control Board process used by NASA to make programmatic decisions does not consistently result in 90-day resolution of these requests.

NASA's partners told us that they need resolution of requests for requirement and certification guidance within 30-90 days in order to avoid delays or additional costs. As of July 2013, over 65 percent of partner requests for alternate requirement standards and deviations have been pending more than 90 days without resolution. Although program officials said they expect to clear most of the requests in 120 days and that this timeline should meet the contractors' needs, we noted a significant number of requests pending more than 120 days. Cost increases and schedule overruns may result if NASA is unable to provide timely and accurate confirmation of requirements and certification guidance.

Coordination of Spaceflight Safety Issues with Other Federal Agencies is Ongoing. The transition from NASA-provided space transportation to low Earth orbit and the ISS to commercially supplied space transportation services, especially for human space travel, requires coordination between multiple government agencies. Although NASA and the Federal Aviation Administration (FAA) have taken positive steps to coordinate their efforts, complex range safety, legal, and insurance issues require additional coordination. NASA and the FAA signed a Memorandum of Understanding in June 2012 on how to collaborate on and coordinate their respective roles;

however, the U.S. Air Force launch ranges are not yet part of what needs to be a coordinated effort and Program officials are attempting to establish a tri-party safety steering group composed of these three agencies. However, the steering group's authority and NASA's authority to influence other agencies' actions has not yet been clearly established. Failure to coordinate effectively with the FAA and the Air Force regarding these issues could adversely affect NASA's efforts to facilitate commercial human space travel.

Management Action

In order to increase the accuracy and transparency of Agency budget submissions, we recommended that NASA revise its guidance, to the extent practical, to ensure that managers of space system development programs in which Space Act Agreements are used provide detailed cost estimates for each year of the program based upon a complete analysis of the program over time before preliminary designs are completed. We also recommended that the Associate Administrator for the Human Exploration and Operations Mission Directorate examine whether more comprehensive costs estimates should be developed by the Commercial Crew Program before the completion of the Critical Design Reviews by NASA's current commercial partners; routinely track adherence to the 90-day goal for responding to contractor requests for alternate requirement standards and variances and explore ways to facilitate the process; and, in conjunction with the FAA and the Air Force, formally establish a tri-agency Safety Steering Group for resolution of issues involving crew and public safety during commercial spaceflight operations.

NASA and the Associate Administrator concurred with our recommendations and agreed to take corrective actions. We consider these proposed corrective actions responsive; therefore, the recommendations are resolved and will be closed upon completion and verification of those actions.

INTRODUCTION

Background

Since the termination of the Space Shuttle Program in July 2011, the United States has lacked a domestic capability to transport crew and – until recently – cargo to and from the International Space Station (ISS or Station).

Consequently, NASA has been relying on the Russian Federal Space Agency (Roscosmos) for crew transportation, and the Europeans and the Japanese for cargo transportation. Between 2012 and 2017, NASA is committed to pay Roscosmos $1.7 billion to ferry its astronauts and international partners to and from the ISS, at prices ranging from $47 million to more than $70 million per seat. After 2017, NASA hopes to secure transportation to the ISS from an American company or companies.

In anticipation of the Shuttle's retirement, Federal law directed NASA to foster the commercial spaceflight industry as a means of developing domestic cargo and crew transportation capabilities to the Station. In November 2005, NASA created the Commercial Crew and Cargo Program Office to work toward this goal and in 2011, activated a separate Commercial Crew Program Office to reflect the increased funding and priority for commercial crew development.

The Office of Inspector General (OIG) issued a report in June 2013 examining NASA's efforts to foster a commercial cargo capability. In that audit, we discussed NASA's funding over the past 7 years of two companies – Space Exploration Technologies Corporation (SpaceX) and Orbital Sciences Corporation (Orbital) – to further development of spaceflight capabilities and, on a separate track, the Agency's contracts with the companies for a combined 20 cargo resupply missions to the ISS.[7] As a complement to that report, this review examines NASA's efforts to foster development of a commercial spaceflight industry capable of carrying NASA astronauts to the Station – known officially as the Commercial Crew Program.

Space Act Agreements

Under the National Aeronautics and Space Act of 1958, NASA has authority to enter into agreements with private companies and other non-Government entities that are not required to follow the general contract requirements of the Federal Acquisition Regulation (FAR). These Space Act Agreements establish a set of legally enforceable commitments between NASA and a partner to accomplish a stated objective related to the Agency's mission that does not involve acquiring specific goods or services. As of September 2013, NASA reported having more than 2,500 Space Act Agreements with companies, educational institutions, state and Federal government organizations, and foreign governments and entities.

Space Act Agreements may be funded, reimbursable, or non-reimbursable (unfunded) agreements. NASA may enter into funded agreements only with domestic partners and only when it cannot achieve its objectives through other types of instruments, including traditional FAR-based contracts.[8] In non-reimbursable agreements, each party bears the cost of participation with no exchange of funds, while in reimbursable agreements a partner reimburses NASA for resources or services the Agency supplies, such as test facilities, supplies, or engineering expertise. Since 2006, NASA has awarded funded Space Act Agreements to 12 commercial partners with the goal of stimulating both the development of commercial capabilities NASA could use to transport its astronauts and cargo to the Station as well as a wider commercial market for spaceflight services.

Using funded Space Act Agreements rather than FAR-based contracts to develop new transportation capabilities has several potential benefits. First, because NASA's partners share development costs and the agreements involve fewer regulations and require less oversight by the Agency, they may reduce the cost to NASA of eventually acquiring services from these companies. Second, because NASA does not impose specific requirements on the companies as part of the agreements, they are free to develop spacecraft designs that will support the needs of both NASA and other potential customers. Third, NASA officials believe that the greater flexibility offered by Space Act Agreements promotes creativity and innovation.

However, NASA's use of Space Act Agreements also poses risks, most prominently limiting the Agency's ability to dictate specific requirements and its oversight of the companies during the development process. These limitations make it more difficult to ensure that the companies will ultimately produce spaceflight systems that meet Agency requirements and that NASA can be confident they will safely carry its astronauts to and from the ISS.

As part of the Space Act Agreements, NASA and the companies agreed to a series of developmental milestones that tie payments to satisfactory completion of such events as design reviews, sub-system testing, and safety and certification reviews. NASA negotiated milestone entry requirements and exit performance criteria with each partner based on their developmental approaches during the solicitation phase of the Agreements.[9] As opposed to a traditional FAR-based contract in which NASA dictates the detailed requirements a company must meet, these milestones and criteria were tailored by the individual companies and negotiated with NASA.

Although NASA has successfully used Space Act Agreements to develop commercial vehicles capable of carrying cargo to the ISS, using this approach to develop major spaceflight systems is still somewhat novel and one that NASA has been refining over time. To ensure a more thorough evaluation process – and based on lessons learned during the commercial cargo program

– the Agency has added more specific criteria for determining when it will consider particular milestones satisfactorily completed. For example, the entry criteria for Commercial Crew's Critical Design Review include partners providing NASA over 40 specific technical products at least 30 days before the event takes place. In contrast, the entry criteria for Critical Design Review for NASA's commercial cargo program called for up to 16 sets of documents due only 7 to 14 days before the review.

Commercial Crew Program

NASA's efforts to facilitate the development of a commercial crew transportation capability began in earnest in February 2010 when the Agency awarded $50 million in Space Act Agreements to five commercial partners to help fund research and design of key technologies and systems. Known as Commercial Crew Development Round 1 (CCDev1), NASA awarded funded agreements to Sierra Nevada Corporation (Sierra Nevada) ($20 million), The Boeing Company (Boeing) ($18 million), United Launch Alliance ($6.7 million), Blue Origin, LLC ($3.7 million), and Paragon Space Development Corporation (Paragon) ($1.44 million) for varying component designs for capsules and launch abort and life support systems. In April 2011, NASA awarded $270 million in Commercial Crew Development Round 2 (CCDev2) Space Act Agreements to Boeing, Sierra Nevada, SpaceX, and Blue Origin for further development ending in Preliminary Design Reviews. Additional milestones were added in September 2011, bringing the total aggregate value of CCDev2 to $315.5 million.[10] Subsequently, NASA entered into additional unfunded agreements with Alliance Techsystems, Inc., Excalibur Almaz, and United Launch Alliance to provide technical assistance on space transportation concepts that included the human rating of various launch systems. (See Appendix B for more information concerning these Commercial Crew Space Act Agreements.)

In December 2011, NASA published the Commercial Crew Transportation 1100 series, which now includes 281 technical and safety requirements commercial spaceflight vehicles will need to meet to receive NASA certification. This documentation also describes the methodology NASA uses to oversee partner activities, namely embedding "Partner Integration Teams" at each company composed of NASA employees who work with company personnel to gain insight and provide technical assistance. According to Program officials, these teams are helping NASA determine whether the partners have met their Space Act Agreement milestones and will be critical when the companies seek certification from the Agency for their vehicles.

Acquisition Strategy

After completion of the initial two rounds of development using Space Act Agreements, NASA planned a two-phased, FAR-based acquisition to further the development of its partners' commercial crew capabilities. Phase 1 was to consist of firm-fixed-price contracts to multiple companies for integrated design and early development followed by a second round of firm-fixed-price contracts for additional development, testing, evaluation, and certification of the contractors' crew transportation systems. Thereafter, NASA planned to enter into a FAR-based contract for crewed missions to the ISS, which it hoped to begin by 2016.

For fiscal year (FY) 2012, NASA requested $850 million for its Commercial Crew Program but received only $397 million. Consequently, NASA revised its Commercial Crew Acquisition Strategy to continue supporting partners' commercial development using Space Act Agreements rather than using FAR-based contracts to procure development work specifically tailored to NASA's requirements. The Agency also delayed the expected first commercial crew flight from the end of 2016 to mid-2017.

Figure 1 depicts NASA's revised acquisition strategy: Space Act Agreements for development (Part 1), FAR-based certification contracts to achieve a crewed demonstration and certification (Part 2), and a FAR-based services contract for delivery of NASA astronauts to the ISS (Part 3). As of August 2013, the same three companies working on development via Space Act Agreements − Boeing, SpaceX, and Sierra Nevada − are simultaneously preparing certification documents for review pursuant to the certification contract. (See Appendix C for a comparison of the original and revised acquisition strategies.)

In Part 1 of its strategy, known as Commercial Crew Integrated Capability (CCiCap), NASA partnered with the companies using Space Act Agreements to mature the design and development of transportation systems that include spacecraft, launch vehicles, and ground and mission systems. Boeing was awarded $480 million, SpaceX $460 million, and Sierra Nevada $227.5 million. In addition to component testing and other design reviews, the goal is for Boeing and SpaceX is to achieve a company-defined Critical Design Review by May 2014 and Sierra Nevada a partial Critical Design Review by October 2013.[11]

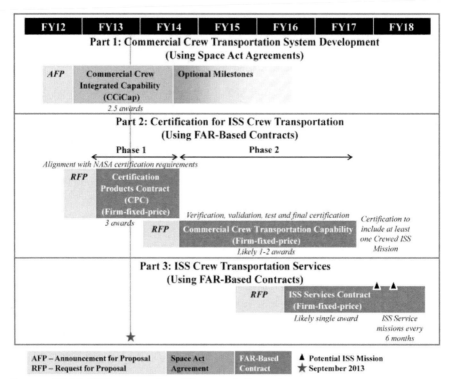

Source: NASA OIG presentation of Program information.

Figure 1. Commercial Crew Acquisition Strategy.

In December 2012, NASA initiated Part 2 of its acquisition strategy by commencing work on the Certification Products Contract (CPC), the first of two phases of certification. These FAR-based contracts seek to ensure the partners' systems will meet NASA's safety and operational requirements. In Phase 1, NASA awarded firm-fixedprice contracts worth approximately $10 million each to Boeing, SpaceX, and Sierra Nevada for delivery and acceptance of certification plans for their crew transportation system. In Phase 2 of certification, known as the Commercial Crew Transportation Capability (CCtCap) and scheduled for award in July 2014, NASA plans to enter into at least one but ideally two firm-fixed-price contracts for the verification, validation, test, and final certification of a crew transportation system or systems. NASA plans to implement Part 3 of the acquisition strategy shown in Figure 1 by issuing a firm-fixedprice ISS transportation services contract to at least one partner in 2017.[12]

Competition in the Commercial Crew Program

NASA's current acquisition strategy attempts to control costs by encouraging competition and using a mix of Space Act Agreements and firm-fixed-price contracts.[13] NASA officials said that entering into Space Act Agreements with multiple partners encourages high performance because the companies can position themselves to compete for the certification and transportation services contract. Officials also said that engaging multiple partners maintains a needed redundancy if one provider cannot perform to the Agency's satisfaction.

While NASA officials said they would prefer to continue to work with at least two companies until the transportation services contract, a lack of funding will likely require them to "down select" to a single partner during Phase 2 of Certification, which is currently scheduled to begin in mid-2014. Moving forward with a single company increases the risk that NASA could be left without a viable commercial option to transport crew to the ISS should issues arise that either significantly delay or render inoperable the selected company's systems. While down selecting to one company is likely to save NASA money in the short term, experience from other major spaceflight programs illustrates that a lack of competition ultimately can drive up costs.[14]

Objectives

Given the importance of the Commercial Crew Program to NASA's human spaceflight efforts, we assessed: (1) Boeing, Space X, and Sierra Nevada's progress toward developing a certified crew transportation capability and (2) the challenges remaining to successful implementation of the Program.

This audit focused on the funded partners in the CCiCap phase of the Commercial Crew Program. See Appendix A for details of the audit's scope and methodology, our review of internal controls, and a list of prior coverage. Prior coverage by NASA OIG includes a June 2011 audit report entitled, "NASA's Challenges Certifying and Acquiring Commercial Crew Transportation Services" (IG-11-022, June 30, 2011). Cost and schedule information in this report is current as of August 31, 2013.

RESULTS

NASA's Commercial Partners Meeting Early Development Milestones but Face Significant Future Challenges

While their development approaches vary, Boeing, SpaceX, and Sierra Nevada are achieving the development milestones specified in their Space Act Agreements with NASA. To date, each partner has completed its initial design and is in the process of finalizing that design and conducting component testing. In December 2012, NASA awarded the three companies Phase 1 FAR-based certification contracts designed to ensure that their spaceflight systems will meet NASA's safety and performance requirements. However, the Critical Design Review and other key developmental milestones historically associated with cost increases and schedule delays lie ahead.

Partners Have Varied Approaches to System Design

Under their Space Act Agreements, each company is required to complete its spacecraft design and a significant amount of component testing by mid-2014.[15] Each company's milestones focus on areas that highlight its individual approach to designing and developing spacecraft. For example, SpaceX's milestones include a Pad Abort Test scheduled for December 2013 and an evaluation of the structural integrity of the Dragon capsule scheduled for January 2014, both of which highlight the company's focus on fabricating components themselves and testing them on a routine basis. Boeing, on the other hand, has extensive experience designing, developing, and building space systems using a traditional acquisition approach for its work on the ISS and the Space Shuttle. Thus, Boeing's milestones focus on component-level design reviews, such as a structural Critical Design Review in October 2013 and a service module propulsion system Critical Design Review in November 2013.

As shown in Table 1, each company has a different approach to spacecraft development. For example, Boeing and Sierra Nevada plan to use the Atlas V launch vehicle to transport their spacecraft to the ISS, while SpaceX will launch the Dragon capsule with its own Falcon 9 rocket. The United Launch Alliance – which is partially owned by Boeing – builds the Atlas V, allowing Boeing to directly influence the upgrades required to certify the launch system for human transportation. In addition, Boeing and SpaceX utilize a similar capsule design that deploys a parachute for landing on a dry surface while

Sierra Nevada plans to land its Dream Chaser on a conventional airport runway.

Table 1. NASA Commercial Partner Profiles

	Boeing	SpaceX	Sierra Nevada
Capsule	CST-100	Dragon	Dream Chaser
Launch Vehicle	Atlas V	Falcon 9	Atlas V
Capability	7 crew or equivalent crew and cargo	7 astronauts	Up to 7 astronauts and cargo
Landing	Dry surface	Dry surface/Ocean	Runway
Total NASA Investment as of August 31, 2013	$415.8 million	$327.8 million	$229.1 million
Total Awarded [a]	$620.9 million	$544.6 million	$363.1 million

[a] Total Awarded includes sum of all funds made available by NASA to these three companies for CCDev1, CCDev2, CCiCap, and CPC.

Source: OIG summary of Commercial Crew Program data.

Partners Are Meeting Early Development Milestones

To date, Boeing, SpaceX, and Sierra Nevada have been meeting the milestones specified in their Space Act Agreements. These include completion of preliminary designs and conducting extensive component-level testing on various aspects of the vehicles. However, the design of the companies' systems have yet to be fully matured or achieve what amounts to a Critical Design Review. This review and the process used to prepare for it often reveal shortcomings that must be addressed prior to finalizing spacecraft design and initiating the manufacturing process.

Past studies have shown that unexpected costs and schedule delays are common during production of space systems.[16] As yet, none of the companies have begun producing an actual spacecraft for a flight to the ISS, only test articles and prototypes.

Table 2 shows each partner's progress and performance as of August 2013. All three have completed a design review similar to the process NASA would use if it were developing the space systems in-house.[17] To date, SpaceX and Boeing have completed all milestones associated with CCDev2 and progressed beyond a Preliminary Design Review level of maturity. Sierra Nevada has one CCDev2 milestone remaining – a test flight of its engineering test article. While SpaceX completed all CCDev2 milestones on schedule, Boeing experienced minor delays with software development and Sierra Nevada experienced significant delays with drop tests.

Table 2. Status of Commercial Crew Program Partners

	2011	2012	2013	2014
	CCDev2 Period (Apr 2011 to July 2012)		CCiCap Base Period (Aug 2012 to Aug 2014)	
Boeing	Boeing completed 16 of 16 milestones from Apr 2011 to Feb 2013 ($112.9 M)			
			Boeing has completed 8 of 20 milestones. ($280.9 M of $480 M)[a]	
SpaceX	SpaceX completed 11 of 11 milestones from Apr 2011 to July 2012. ($75 M)			
			SpaceX has completed 7 of 16 milestones. ($249 M of $460 M)[a]	
Sierra Nevada	Sierra Nevada has completed 14 of 15 milestones. The last milestone was expected to be complete by Oct 2013. ($97.6 M of $105.6 M)			
			Sierra Nevada has completed 4 of 12 milestones. ($107.5 M of $227.5 M)[a]	

[a] Milestone completion as of August 31, 2013. Dollars are in millions.
Source: OIG summary of Commercial Crew Program data.

In addition, as of August 2013, all three partners were meeting their CCiCap milestones and have concurrently begun work on their CPCs. CPC Phase 1 work, which includes submitting reports and plans to NASA to ensure compliance with Agency requirements and certification guidance, began in January 2013.[18] By May 2014, each partner is scheduled to complete CPC Phase 1, while Boeing and SpaceX should complete a Critical Design Review. Sierra Nevada – which received about half the amount of money NASA provided the other two partners – is aiming to achieve the first of multiple incremental Critical Design Reviews by October 2013.

Boeing. As of August 2013, Boeing has completed 8 of their 20 CCiCap milestones and is scheduled to receive up to $480 million. In addition, NASA has the option to execute additional milestones designed to further mature Boeing's spacecraft design and enable a full system demonstration.

- Milestone 1 (August 2012) – completed an integrated system review.[19]
- Milestone 2 (October 2012) – completed a production design review to determine if subsystems are sufficiently mature to enter production and production plans are correct.
- Milestone 3 (November 2012) – completed a phase 1 safety review of requirements, vehicle architecture and design, and associated safety products.
- Milestone 4 (January 2013) – completed a software integrated engineering release demonstrating the capabilities of guidance and navigation and control software.
- Milestone 5 (January 2013) – completed a landing, recovery, and ground communications design review.
- Milestone 6 (February 2013) – completed a design review of the launch vehicle adapter by which the CST-100 capsule will be attached to the Atlas V launch vehicle.
- Milestone 7 (April 2013) – completed an integrated wind tunnel test to analyze the aerodynamic attributes of the launch vehicle and capsule during ascent.
- Milestone 8 (May 2013) – completed a dual engine centaur liquid oxygen duct development test to assess interface and water flow for liquid oxygen ducts.

Moving forward, Boeing has scheduled other reviews and testing expected to culminate in a Critical Design Review in April 2014. During CCDev2, Boeing experienced delays in software development and completed its final CCDev2 milestone in February 2013, 7 months behind schedule. According to NASA officials, Boeing must continue to address software issues or risk delaying its plans for a 2016 demonstration flight to the ISS.

SpaceX.[20] As of August 2013, SpaceX has completed 7 of its 16 CCiCap milestones and is scheduled to receive $460 million. In addition, NASA may execute optional milestones designed to progress the company's system to a full demonstration by December 2015.

- Milestone 1 (August 2012) – held a kick-off meeting to present NASA with the company's plans for completing CCiCap milestones.
- Milestone 2 (August 2012) – submitted an analysis of its finances for its crew program to demonstrate the company's ability to successfully complete CCiCap goals.
- Milestone 3 (October 2012) – completed an integrated systems requirements review, including a review of its plans for system design, production, and operations and how they will meet NASA's requirements.
- Milestone 4 (December 2012) – completed a ground systems and ascent Preliminary Design Review to demonstrate that its system meets all related requirements with acceptable risk and within schedule constraints.
- Milestone 5 (March 2013) – completed a pad abort test review to demonstrate the maturity of the pad abort test article design and test concept of operations.
- Milestone 6 (May 2013) – completed a human certification plan review to define in detail the company's approach to certifying the design of the spacecraft, launch vehicle, and ground and mission operations systems.
- Milestone 7 (July 2013) – completed the on-orbit and entry Preliminary Design Review to demonstrate that its system can orbit, rendezvous, and dock with the ISS.

Moving forward, SpaceX has scheduled additional testing and certification milestones, including an on-orbit and entry Preliminary Design Review and an in-flight abort test review as they work toward a Critical Design Review in March 2014.

Sierra Nevada. As of August 2013, Sierra Nevada has completed 4 of their 12 CCiCap milestones and is scheduled to receive up to $227.5 million. The company has laid out a plan that includes advanced development milestones culminating in a full systems demonstration to the ISS before the end of 2017. Unlike the other partners, NASA has only funded Sierra Nevada for the first of multiple incremental Critical Design Reviews. However, NASA may choose to fund additional incremental Critical Design Reviews and other optional milestones.

- Milestone 1 (August 2012) – completed a review of its plan for implementing CCiCap.
- Milestone 2 (October 2012) – completed an integration system baseline review demonstrating maturity of its integration plan and operations design.
- Milestone 3 (January 2013) – completed an integration system safety analysis review demonstrating the safety of the Dream Chaser Space System.
- Milestone 5 (June 2013) – completed an investment milestone reflecting its financial commitment and ability to co-fund its system.[21]

Moving forward, Sierra Nevada has scheduled additional reviews and tests, including wind tunnel testing scheduled to be completed in February 2014. Sierra Nevada faced challenges during CC Dev2 with respect to performing a successful drop test and has yet to complete this milestone, which was scheduled for October 2013. To maintain its schedule during CCiCap, Sierra Nevada must work to address these delays and complete the remaining milestones, which consist of extensive testing of crew and life support systems and additional aerodynamic tests.

NASA Must Overcome Major Hurdles to Provide Commercial Crew Transportation on Current Schedule

Although NASA's three commercial crew partners have been achieving their early development milestones, the Commercial Crew Program faces significant internal and external challenges to meeting NASA's goal of securing crew transportation to the ISS by 2017. First, for the past several years the Program has received significantly less funding than NASA requested. As a result, NASA has delayed the target date for the first crewed flight to the ISS to FY 2017, only 3 years before the currently scheduled end of ISS operations.[22] Further, experience has shown that reduced funding profiles are not consistent with best practices for space system development and continuation of this funding pattern could result in additional cost increases, schedule delays, and a lack of competition later in the Program's development. Second, NASA has yet to project the total amount of funding required by year, which makes it difficult for the Agency to manage its wider portfolio of spaceflight programs and at the same time reduces the transparency of the Program's budget submissions. Third, the process for providing timely guidance to partners for satisfying NASA's human rating and certification requirements could be improved. If NASA is unable to confirm design requirements and provide certification guidance in a timely manner, the companies could face costly and time-consuming redesign work late in system development. Finally, coordination of important safety issues with the Federal Aviation Administration (FAA) and the Air Force is ongoing, but not yet resolved. Failure to address these challenges could significantly delay the availability of commercial crew transportation services and extend U.S. reliance on the Russians for crew transportation to the ISS.

Funding Instability Jeopardizes NASA's Plans to Begin Certified Crew Missions to the ISS by 2017

Throughout its existence, NASA's Commercial Crew Program has received budget appropriations that were substantially lower than requested. Perhaps the most tangible impact of these funding shortfalls has been the 2-year delay of the first crewed demonstration flight to the ISS by a commercial partner.

Specifically, for FY 2013, the Program received $525 million or 37 percent less than the President's budget request. The year before, the Commercial Crew Program requested $850 million but received $397 million, less than half its request. Overall, when comparing actual funds received to the original FY 2011 budget request, which was based on a $6 billion program developed in 5 years, the Program has received 38 percent of its requested funding for FY 2011 through FY 2013 (see Figure 2).

For subsequent budget requests beginning in FY 2012, the Office of Management and Budget reduced the Program's annual appropriations request and created a "flat-line" budget profile by spreading funding evenly over subsequent fiscal years (see Table 3). Specifically, for FY 2013 Commercial Crew Program managers had to revise program schedules after their budget was reduced from the $830 million requested by the President to the $525 million appropriated by Congress.[23] Generally speaking, we determined that each year's budget decrement has resulted in an additional year of schedule delay. Even if the Program receives its full budget request in future years, the cumulative difference between the Program's initial budget requests and receipts over the life of the Program would be approximately $1.1 billion.

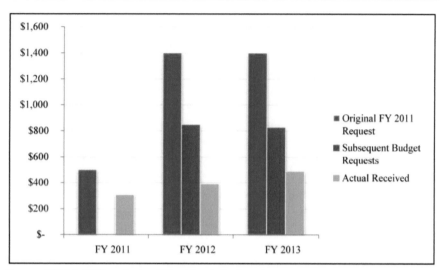

Source: NASA OIG analysis of President's budget data.

Figure 2. Evolution of Commercial Crew President's Budget Requests from FYs 2011 through 2013 (Dollars in Millions).

Table 3. Commercial Crew Program Budget Requests by FY (Dollars in millions)

	FY09	FY10	FY11	FY12	FY13	FY14	FY15	FY16	FY17	FY18	Total
FY09 Budget Request	$0	$0	$0	$0	$0						$0
FY10 Budget Request	$51	$0	$0	$0	$0	$0					$51
FY11 Budget Request	$51	$0	$500	$1,400	$1,400	$1,300	$1,200				$5,851
FY12 Budget Request	$51	$0	$321	$850	$850	$850	$850	$850			$4,623
FY13 Budget Request	$51	$0	$321	$397	$830	$830	$830	$830	$830		$4,920
FY14 Budget Request	$51	$0	$321	$397	$525	$821	$821	$821	$590	$371	$4,719
Total Shortfall	$0	$0	($179)	($1,003)	($875)	($479)	($379)	$821	$590	$371	($1,133)

*Prior years' actual received amounts are in grey. Total Shortfall amounts are based on actual received amounts versus the original FY11 request where appicable. Numbers in parentheses are negative.

Source: NASA OIG analysis of President's budget data.

Funding Shortfalls Led to Extended Use of Space Act Agreements and Delays to the First Crewed Launch. NASA's original budget estimate for the Commercial Crew Program projected that $1.4 billion would be required for FY 2012. Ultimately, NASA requested $850 million for the Program that year but received just $397 million.

NASA's first acquisition plan for developing the Commercial Crew Program anticipated the use of FAR-based contracts starting in late FY 2012 for the integration phase of development. During integration, NASA expected its partners to progress to a point where their system designs were mature. According to NASA officials, FY 2012 funding was insufficient to execute this plan. As a result, they continued to use funded Space Act Agreements to support the companies' development efforts.[24] Under this revised acquisition plan, NASA is also using FAR-based contracts to start certification activities and running that process parallel to activities under the Space Act Agreements. Even with this shift in procurement strategy, the budget reductions resulted in schedule delays, postponement of certification, and a delay in the completion of commercial crew development from FY 2016 to mid-FY 2017.

NASA officials told us that keeping multiple partners in the Program for as long as possible is essential to maintain redundant capability and competitiveness. For the next phase of development, including full system certification and demonstrations to the ISS, NASA plans to use a FAR-based contract.

However, continued funding shortfalls will only increase pressure on NASA to "down select" to a single company. Acknowledging this point in a March 2013 hearing, the NASA Administrator stated that funding profiles will affect the number of companies NASA will be able to fund during the final certification phase.[25]

Although NASA has considered the possibility of continuing to use Space Act Agreements during the final phase of development, as noted earlier in this report, Agency officials told us they cannot impose specific requirements on partners under a Space Act Agreement in the same way they can with a FAR-based contract. Accordingly, they do not believe that Space Act Agreements would be suitable vehicles for certifying that particular systems meet NASA safety and human rating requirements.

Current Funding Plan Increases Risk of Further Cost Overruns, Schedule Delays, and Impediments to Competition. Decreased and flat-lined funding has already caused a 2-year delay in commercial crew development efforts. Moreover, the Program's anticipated funding for FY 2014 and beyond is not consistent with best acquisition practices and could manifest in additional problems down the road. Prior audit work by the GAO has shown that the most effective budget profile for large and complex space system development programs is steady funding in the early stages of development and increased funding during middle stages.[26] An absence of sufficient funding early in development increases cost risks and the potential for schedule delays because project managers have to defer development of critical technologies to a time when integration of those technologies may be more difficult or when the costs of material and labor may be greater. Shifting tasks to later project phases also results in higher fixed costs given the need to sustain a work force longer than originally planned or to add shifts in an attempt to make up for lost time. In addition, insufficient early funding for development decreases management's ability to identify and address key risks. Finally, insufficient funding may make it difficult to maintain two partners thereby eliminating the competition. Agency officials maintain that competition increases partner performance and helps drive down the price of future contracts.

A typical space system life cycle model shows a bell-shaped funding curve for research, development, testing, and evaluation because more resources are needed as development progresses and programmatic risks are identified and remediated (see Figure 3). However, while NASA's FY 2011 budget request for its Commercial Crew Program closely resembled the traditional bell curve and anticipated completing development in FY 2015, the Program's actual funding profile for FY 2012 and FY 2013 was flattened and the development schedule was extended. The FY 2014 budget request again reflects a bell curve, with development funding extended through FY 2018.

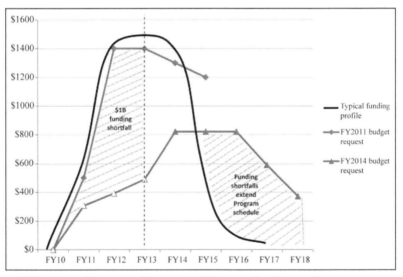

Note: Solid white milestones reflect actual dollars received in previous years.
Source: NASA OIG analysis of President's Budget data and extrapolation to guidance
    from GAO's Cost Estimating and Assessment Guide.

Figure 3. Program Evolution of Developmental Funding Budget Requests vs. Typical Funding Profile (Dollars in Millions).

If the shortfall in funding continues, NASA officials said it will be difficult for the Program to meet its goal of completing development by 2017. While the commercial partners could increase their contributions to offset the Government's reduced funding and help keep the Program on track, representatives from one company expressed uncertainty with regard to the continued development of their spacecraft in the absence of additional funding from the Agency. On average, the three Commercial Crew partners are contributing under 20 percent of the CCiCap development costs for their spaceflight systems.[27]

Two key advisory groups – the Aerospace Safety Advisory Panel (ASAP) and the National Research Council – recently published reports addressing the effects of funding shortfalls. In its 2012 Annual Report, the ASAP stated:

From its inception, this program [Commercial Crew] has been funded at levels far below what would be expected for a traditional program. NASA's ability to successfully complete the certification process as currently planned and to begin flying [missions] by FY 17 will require increased funding for the program starting in FY 13... the ASAP is concerned about what tradeoffs NASA and the [partners] will take to accommodate future budget shortfall. There appear to be few options available to NASA when the budget is reduced. These include stretching the schedule, reducing performance, and/or obtaining additional funding. Given NASA's budget history, it is unlikely there will be additional funding.[28]

The National Research Council made asimilar observation in its 2012 report when addressing NASA programs in general:

> ...there has been some instability at the programmatic level and the out-year projections in the President's budget are unreliable, which makes it difficult for program managers to plan activities that require multi -year planning. Numerous times the agency initiated new programs with the expectation that budgets would increase to support them, only to have no increases emerge...The approach to and pace of a number of NASA's programs, projects, and activities will not be sustainable if the NASA budget remains flat, as currently projected.[29]

Aligning Cost Estimates with Program Schedule May Enhance Management and Improve Transparency of Commercial Crew Program

According to NASA guidance, managers should develop year-by-year life cycle cost estimates for their programs before completing Preliminary Design Review.[30] A life cycle cost estimate refines a program's overall cost estimate by determining all possible costs required to develop, produce, deploy, and sustain a particular program and projecting them over time (by year). This policy is designed to prevent development delays in future years caused by inadequate funding levels.

NASA's Chief Engineer told us that NASA policies requiring life cycle cost estimates do not apply to development efforts that use Space Act Agreements, and that developing life cycle estimates for such programs has proven challenging. Therefore, despite completion of Preliminary Design Review by NASA's commercial crew partners, Agency officials have yet to develop a life cycle cost estimate for the Program. Instead, NASA developed independent government cost estimates that project the overall costs to complete development and certification.[31] An independent government cost estimate simply documents the government's assessment of the program's most probable overall cost without aligning the cost estimates to a schedule. The Chief Engineer did state that once a FAR-based contract is issued for CCtCap, which occurs around the same time the partners are expected to complete their Critical Design Reviews, the Program will become more traditional and a life cycle cost estimate will be feasible.

Because partnering with commercial companies through Space Act Agreements is a novel approach to developing human spaceflight transportation capabilities, NASA was faced with an array of challenges in developing cost estimates for the Commercial Crew Program. For example, Program officials noted that because of varying design approaches, each partner's estimated costs are different, making it difficult to complete a comprehensive life cycle cost estimate. In addition, NASA space system acquisition programs typically use the NASA/Air Force Cost Model to produce a life cycle estimate, which relies on historical information about NASA and Air Force space projects. However, because of the unique development approach used for this Program, little comparable historical data exists. Adding to the difficulty is that under a Space Act Agreement, developmental costs covered by the companies are not a part of Program costs and are generally proprietary. Given these factors, cost estimators could not fully rely on the model's early estimates.

We acknowledge the difficulties in establishing a comprehensive life cycle cost estimate for a program using a development approach that is evolving and for which limited comparable historical data exists. However, without a detailed cost estimate for each year of the program based upon a complete analysis of each subcomponent over time, it is difficult for NASA to calculate how much funding is required each year given that costs over time can fluctuate significantly. Furthermore, the lack of a comprehensive cost estimate makes it more difficult for NASA to manage its wider portfolio of spaceflight programs, particularly in a constrained budget environment.

Improved guidance requiring detailed cost estimates aligned with the program's schedule over time could help enhance the transparency of NASA's budget submissions. To this point, members of Congress and the ASAP have stated that in the absence of this type of information it is difficult to know with any level of confidence the amount of government money required to develop a viable commercial crew transportation capability.[32]

In early 2013, the private consulting group Booz Allen Hamilton (Booz Allen) completed an independent assessment of the Commercial Crew Program's government estimate that included costs from the end of CCDev2 through certification.[33] Overall, Booz Allen found that the Program's cost estimates were of high quality and followed standard cost estimating best practices. However, the assessment found that the estimates were optimistic, and that the Program was likely to experience cost growth. In addition, Booz Allen noted that without costs projected over the life of the Program, NASA officials will not be able to independently evaluate each partner's progress. Booz Allen recommended several efforts to improve the accuracy of the cost estimates, including using an integrated schedule and maintaining risk register data.[34]

In our judgment, NASA is now better positioned to develop improved guidance for cost estimating when systems are developed using Space Act Agreements in light of historical cost data received under the Commercial Crew Program.

Timeliness of NASA's Process for Considering Alternative Human Rating Requirements Remains a Concern

NASA's current process for considering partners' requests to use alternate standards or to deviate from NASA's human rating and certification requirements frequently does not result in timely decisions.[35] The Agency's current process was developed to ensure NASA management sufficiently considers dissenting opinions on safety-related issues. While we acknowledge the importance of considering all opinions on safety issues, an inability to address partners' requests in a timely manner risks costly redesign work or delays to the Program schedule. Although Program officials expected to process the majority of the partners' requests by mid-September 2013, the partners are expected to submit additional requests before the certification planning is completed.

Human Rating Requirements

"Human rating" is the process of assuring that a spacecraft or launch vehicle is capable of safely transporting human beings. Human rating concepts developed over the past 60 years include:

- using simpler designs to avoid complex components,
- using well-established and proven aerospace design standards,
- incorporating sufficient redundancy in all critical systems, and
- avoiding untried or unproven technology.

According to NASA, a human rated system must accommodate human needs; effectively utilize human capabilities; control hazards; manage safety risks; and, to the maximum extent possible, provide the capability to safely recover the crew from hazardous situations.[36] Compliance with these requirements leads to certification that the system is suitable for crewed spaceflight. Before securing transportation for its astronauts on a commercial spaceflight system, the Agency needs to review and approve the system to ensure it meets NASA's human rating requirements.

The OIG, GAO, and others have expressed concerns about NASA's approach to determining baseline human rating and safety requirements for commercial crew vehicles, and for ensuring that contractors adhere to those requirements. In order to stimulate innovative designs, NASA intentionally did not impose human rating and safety requirements on its commercial partners during early development. While such an approach may foster innovation, it also can raise concerns that partners may be developing systems that NASA will ultimately be unable to certify.

To address this challenge, in December 2011, NASA published a series of detailed requirements, certification, and management documents that provide its commercial partners with the Agency's specific safety and human rating objectives:

- Crew Transportation Plan (CCT-PLN-1100) defines the processes for Crew Transportation System certification and flight readiness.
- System Design Reference Missions (CCT-DRM-1110) defines reference missions to the ISS.
- Crew Transportation Technical Management Processes (CCT-PLN-1120) provides the technical management processes that support the certification effort.

- ISS Crew Transportation and Services Requirements Document (CCT-REQ-1130) establishes requirements that will be met to achieve certification to transport NASA crew.
- Crew Transportation Technical Standards and Design Evaluation Criteria (CCT-STD-1140) and Crew Transportation Operations Standards (CCT-STD-1150) are intended to guide the development of the crew transportation system.[37]

These documents outline the fundamental elements a system must satisfy to receive certification from NASA. According to NASA and partner personnel, the commercial partners have used these documents to ensure that they are integrating NASA's requirements into their transportation system designs. However, NASA has not yet certified whether any of the partners' systems meet the Agency's requirements.

Figure 4 depicts NASA's strategy for ensuring the partners meet the Agency's human rating and safety requirements. In December 2012, NASA awarded CPC – the first of a two-phased, FAR-based contract under which the Agency plans to review the partners' transportation system designs to ensure they meet its safety and operational requirements.[38] In CPC, the partners provide products considered critical to establishing expectations for certification by NASA. The CPC statement of work provides criteria and requirements to guide the contractors' efforts toward NASA certification. These products include the following deliverables:

- *Alternate Standards* – The Contractor will deliver alternate standards proposed to satisfy the NASA "meet the intent of" standard requirements.
- *Hazard Reports* – The Contractor will deliver hazard reports for at least catastrophic hazards. The hazard reports will address all phases of the mission to the ISS. Hazard reports will include descriptions, effects, risk assessments and actions necessary to eliminate or control the hazard.
- *Verification and Validation Plan* – The Contractor will deliver a Verification and Validation Plan. The Plan will identify the verification and validation activities, methods, deliverables, and processes that result in objective evidence that all elements of the design, production, and operation meet the performance requirements and accomplish the intended design reference mission to the ISS when operated in the intended environment. The Contractor should deliver

any proposed variances to meeting NASA's certification requirements.

- *Certification Plan* – The Contractor shall deliver a Certification Plan that will define an integrated strategy for certification of the complete system and will clearly define the order of execution and schedule.

During CCtCap, scheduled to start in the middle of 2014, NASA plans to enter into one or two firm-fixed-price contracts pursuant to which the Agency will endorse the partner's own internal certification in order to grant final certification of a crew transportation system or systems. This strategy anticipates the Agency certifying at least one operational crew transportation system in time for missions to the ISS by 2017.

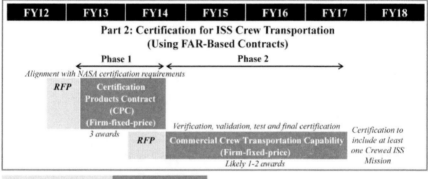

Source: Commercial Crew Program Acquisition Strategy.

Figure 4. Certification Strategy.

Process to Resolve Partner Requests for Alternate Standards and Requirements Variances. Given the partner's varying system designs, the Program has developed a process for assessing requests to use alternate standards and to vary from its stated requirements with a 90-day goal for NASA to respond to such requests. However, in the past, the process has taken as much as 9 months, and according to NASA officials, for other Agency programs such as the Ares I, even longer. The Commercial Crew Program office has acknowledged that this process can be lengthy and identified it as a significant risk to the Program. However, we found that the Program Office is not adequately tracking the timeliness of responses to partner requests.

Requests related to Commercial Crew Program requirements are reviewed by the Commercial Crew Program Control Board while requests related to ISS requirements go to the Space Station Program Control Board. Requests can go to both boards if an integration issue exists.

Technical or safety requests of either type are first directed to a Technical Review Board for consideration and this Board makes recommendations to the full Program Control Board. Throughout the process, supporting organizations can provide dissenting opinions, which require additional analysis.[39] If a particular organization is not satisfied with the Program Control Board's decision, the issue can be elevated to a higher authority, including all the way to the NASA Administrator for final determination.

Figure 5 depicts the general process used by both Boards.

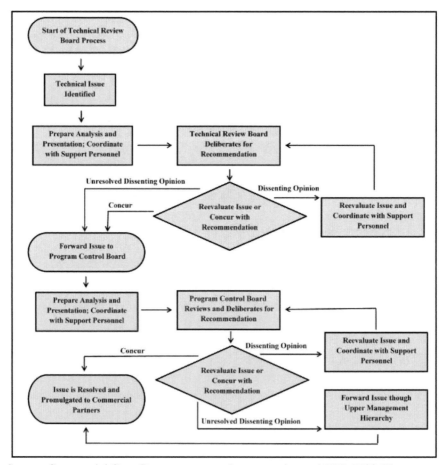

Source: Commercial Crew Program process documentation and NPR 7120.5E.

Figure 5. Technical Review and Program Control Boards Process.

This process was developed to ensure NASA management considers safety-related dissenting opinions when reaching a consensus, and officials said the process may take longer when the specific issue involves more risk to the mission. For example, in the original iteration of its human rating requirements, the Agency did not require pressure suits be tested to ensure they maintain a living environment.[40] However, in November 2011, NASA personnel questioned this omission, which resulted in the issue being forwarded to the Technical Review Board and eventually to the full Program Control Board for review. During the review, dissenting opinions were elevated to the Program Control Board and senior NASA officials were consulted. In July 2012, NASA clarified the requirement and required the pressure suits be tested to ensure they maintain a living environment. According to Program officials, the 9 months it took to resolve this issue is not atypical for resolution of high-risk issues associated with human spaceflight programs.

As of July 2013, NASA has received 86 alternate standards or variances requests from its commercial crew partners. Of the requests that have been in the review process for more than 90-days, over 65 percent remain unresolved. According to Program personnel, part of the delay is attributable to the partners not providing complete information, requiring the Program Office to request additional data.

The Commercial Crew Program Office has acknowledged that the process for addressing requests for alternate standards or variances has been lengthy in the past. However, Program officials anticipated speeding up the pace of processing and completing the majority of the requests by mid-September 2013. Program officials also said that verbal feedback on the request is shared with the Partners long before the requests are formally closed.

We acknowledge the recent efforts the Agency has made in more timely processing requests for alternate standards or variances. Nonetheless, because the multi-step review process historically has been time consuming, and the partners are expected to submit additional requests before the certification planning is completed, resolving the current and any new requests in a timely manner is critical to Program success. For example, many of the program replies to the partner requests involve asking for more information – after a 90-120 day period has already passed.

In our judgment, greater visibility regarding the timeliness of NASA responses during program briefings would help prioritize and emphasize the importance of bringing resolution to the partner requests for alternate standards and variances within the 90-day goal. We found that although the

Program Office regularly briefs NASA management on the progress of the partner requests, the briefings did not include a summary of the timeliness of NASA's responses, nor did they identify the number of requests over 90 days.

We acknowledge the need for the Agency to ensure that safety-related concerns are comprehensively considered. However, not resolving alternate standards and variance requests in a timely manner could adversely affect the Program. Specifically, partners may delay production of their final designs or move forward with requirements that are not fully defined. Previous reports by the OIG, GAO, and others have shown that requirement changes late in product development can lead to design rework that is both costly and time consuming.

Coordination of Spaceflight Safety Issues with Other Federal Agencies is Ongoing

NASA's foray into securing commercial transportation for its astronauts requires a greater level of coordination between the Agency and the FAA than has been the case with its previous crewed and uncrewed spaceflight missions. Although the two agencies have taken positive steps to coordinate their efforts, complex range safety, legal, and insurance issues remain unresolved.

NASA and FAA Responsibilities for Commercial Spaceflight Activities. For more than 50 years, NASA was solely responsible for the safe conduct of its space launches. However, because the Government will not own or control the operation of the space systems developed by NASA's commercial crew partners, their launch and reentry are deemed commercial activities subject to the jurisdiction of the FAA.[41]

The Commercial Space Launch Amendments Act of 2004 temporarily restricts the FAA's authority to regulate the commercial space industry. Specifically, until 2015, the FAA's authority is limited to regulating launches and reentries, and therefore at present the FAA does not license commercial spacecraft, requires no airworthiness certificate for these companies, and imposes no crew escape requirements during launch. Beginning in October 2015, the FAA is authorized to promulgate regulations to ensure crew and participant safety.[42] Eventually, oversight of the commercial spaceflight industry will parallel the way in which the FAA approaches airline safety, although NASA will continue to retain responsibility for ensuring the safety of its crews, workforce, and infrastructure during commercial launches carrying NASA astronauts.[43]

Because of these overlapping authorities, both NASA and the FAA must coordinate their efforts to avoid imposing conflicting requirements on commercial spaceflight operators. The NASA partner companies we spoke with said they are concerned about the impact conflicting requirements could have on their operations. Officials from one company said it is important that the FAA publish its regulations as soon as possible to enable each company in the market to operate under the safest conditions. They further explained that common sense regulation could help prevent an early accident that could result in over regulation – which could have a costly and negative effect on the fledgling industry. Other partners said they would prefer that the FAA postpone its regulations until both the industry and NASA requirements are fully developed because they do not want competing, overlapping, or contradictory requirements from NASA and the FAA.

Coordination Efforts are Progressing. As discussed in our 2011 audit report, NASA and the FAA took initial steps to coordinate their efforts by having discussions and co-locating personnel at various NASA and FAA offices.[44] Since that time, NASA and the FAA have taken additional steps to coordinate their efforts. For example, in June 2012 the two agencies signed a Memorandum of Understanding agreeing to collaborate on several efforts, including:

- providing a stable framework for the U.S. space launch industry,
- avoiding conflicting requirements and multiple sets of standards,
- advancing both public safety and crew safety, and
- advancing the interests of NASA-certified U.S. commercial launch operators responsible for transporting U.S. and partner astronauts to the ISS.

NASA and the FAA are working to both achieve consistency between NASA and non-NASA flight activities and to avoid conflicts between FAA regulations and the Commercial Crew Program requirements. To that end, the agencies have developed a list of items requiring coordination, such as the authority to approve changes while in flight, including designating emergency landing sites. Although this list is still evolving, as of August 2013, the agencies have resolved 24 of 59 issues and are working to settle the remaining issues. However, many of the outstanding items are among the more complex issues facing the two agencies and, as the Commercial Crew Program has acknowledged, will require considerable technical and legal review as well as the agreement of both NASA's Commercial Crew Program Control Board and the FAA Office of Commercial Space Transportation.

Examples of these complex issues include agreed-upon credentials for pilots and crew, financial responsibility for claims, and the claims waiver requirements.[45] Commercial Crew Program managers noted that NASA, the FAA, and the U.S. Air Force – which operates the launch ranges most often used by the commercial partners – must develop a process that encourages uniformity among Government oversight authorities in the implementation of requirements.

Mechanism Lacking for Official Resolution of Joint Issues. One hindrance to ongoing coordination between NASA and the FAA is the lack of an official mechanism to address joint issues and obtain documented resolution. NASA staff has acknowledged that documented decision-making is necessary for maintaining internal control of the resolution process. However, the process for documenting issue resolution through the Commercial Crew Program Control Board and the FAA Office of Commercial Space Transportation is time consuming and not firmly established. For example, the FAA participates in meetings of the Commercial Crew Program Control Board only as an observer/liaison and FAA representatives have indicated they do not view Board deliberations as binding to the FAA. Therefore, NASA staff members coordinating resolution of requirements issues currently record the decisions from the Commercial Crew Program Control Board in addition to documentation of agreement of those decisions by the FAA Office of Commercial Space Transportation.

Furthermore, consistent application of safety principles is necessary so that crew safety is held to the same standard for Government or commercially-licensed activities. The Commercial Crew Program has initiated formation of a tri-party safety steering group involving NASA, FAA, and U.S. Air Force staff.

The group will be chartered to promote consistency and uniformity among the various Government entities with whom commercial providers must work to provide commercial crew space transportation. As of August 2013, the "Commercial Human Spaceflight Launch and Re-Entry Steering Group" (steering group) had developed a draft charter. However, the steering group's authority and reporting structure remains to be established. Apart from the steering group's work, Commercial Crew Program officials have proposed developing a Commercial Crew Program/FAA Program Management Plan to document joint processes and information exchange for the execution of licensed missions.

Source: NASA.

Figure 6. Multiple Agencies Routinely Involved in Commercial Spaceflight: SpaceX's Falcon 9 with Dragon Capsule Launches from Cape Canaveral Air Force Station En Route to the ISS on a NASA Resupply Mission.

Regulatory Issues Remain that Could Impact Mission Schedules. Without a formal process to resolve complex coordination issues, it will be difficult for NASA to influence the actions of the other Government entities that have a role in commercial crew spaceflight. Of greatest concern are issues involving safety, such as whether a change in a spacecraft's reentry trajectory would violate the spacecraft operator's license. An extreme example is a situation in which the weather is deteriorating and the FAA's license for the flight (provided 180 days in advance of the flight) may not allow for a landing at a place or time that would provide the crew with the best chance of a safe return. In such an event, proper FAA guidance and range responses would be critical to ensuring crew safety. Although air traffic controllers make these types of decisions routinely for changes in airline flights, a similar procedure does not currently exist for commercial space launches.

Other unresolved coordination matters involve legal and liability issues. These include:

- FAA licenses cover launch and reentry, but the definition of the "end of launch" and "beginning of reentry" for Commercial Crew Program missions remains to be resolved. For instance, end of launch could be considered when rocket stages separate, or in the case of expendable launch vehicles, the "last exercise of control" over the launch vehicle, which historically occurs with the venting of the tanks and discharging of batteries. A failure to define each term could result in confusion over which agency has control or decision making authority during flight.

- A question exists regarding the legality of NASA astronauts operating commercial vehicles given the statutory definition of "crew" as employees of the licensee.

- Current FAA regulations parallel U.S. Air Force range requirements. For past-crewed missions, NASA negotiated with the range for implementation of contingencies that specifically addressed the safety of crews, including procedures for emergency egress and landing. For FAA-licensed launches, NASA has no authority or official avenue to influence range safety policy or implementation.

Conclusion

NASA continues to make progress in commercial spaceflight development, as recently demonstrated by SpaceX's successful cargo resupply missions to the ISS and Orbital's maiden test flight of its cargo delivery rocket. However, the ability of commercial partners to successfully develop safe, reliable, and cost effective space transportation for human transportation to the ISS is far more complex and unproven.

Reduced budgets for NASA's Commercial Crew Program have forced the Agency to delay completion of the Program's development phase until 2017, and if this funding trend continues the risk of delay beyond 2017 increases significantly. As such, it is crucial that the Agency make a concerted effort to develop comprehensive and timely cost estimates for its Commercial Crew Program. NASA's publication of commercial crew requirements and certification documents was a major step forward in the evolution of the Program, but the path to certifying a commercial partner's spaceflight system remains uncertain. Given that each partner has a varied business and technical approach to development, it is important that NASA utilize a process that both ensures safety-related concerns are sufficiently examined and that deviations, exceptions, and waivers are processed in a timely manner. Failure by NASA to confirm design requirements and provide timely certification guidance regarding deviations, exceptions, and waivers could impact contractors' ability to move forward with production of their spaceflight systems.

Finally, NASA's ability to obtain adequate funding from Congress and coordinate effectively with the FAA and Air Force to manage the commercial spaceflight industry are key factors underpinning the success of the Agency's crewed commercial spaceflight efforts. Failure to secure these capabilities will require NASA to negotiate for additional seats on the Soyuz, which by 2017 will cost over $70 million each. Moreover, given that the operational status of the ISS beyond 2020 is yet to be determined, delays beyond 2017 may call into question the viability of the Program.

Recommendations, Management's Response, and Evaluation of Management's Response

Recommendation 1. We recommended that NASA, to the extent practical, revise guidance to ensure that managers of space system programs developed using Space Act Agreements provide detailed cost estimates for each year of the program based upon a complete analysis of the program over time before preliminary designs are completed in order to increase the accuracy and transparency of Agency budget submissions.

Management's Response. NASA concurred with our recommendation, noting that it will review existing policies and guidelines to determine the appropriate mechanism for satisfying the intent of this recommendation. NASA stated that while it will be extremely challenging to prepare detailed cost estimates for each year based upon a complete analysis before preliminary designs are completed, the Agency will endeavor to increase the accuracy and transparency of Agency budget submissions for future use of funded Space Act Agreements. NASA expects to complete any required updates to its existing policies and guidelines by mid-calendar year 2014.

Evaluation of Management's Response. We consider NASA's proposed corrective actions responsive; therefore, the recommendation is resolved and will be closed upon completion and verification of the corrective actions.

Recommendation 2. We recommended that the Associate Administrator for the Human Exploration and Operations Mission Directorate examine whether more comprehensive costs estimates should be developed by the Commercial Crew Program before completion of the Critical Design Review.

Management's Response. The Associate Administrator concurred with our recommendation, noting that the Program currently employs a series of reviews involving both internal and external checks that continually examine the Commercial Crew Program's status and that these reviews can identify cost risks or other indicators that may suggest the need for more comprehensive cost estimates. The Associate Administrator stated that NASA will take additional actions to address our recommendation, including collaborating with the ISS Program to examine service costs and updating the Independent Government Cost Estimates for both the FY 2016-2020 budget cycle and after the CCtCap award. The Associate Administrator stated that all additional actions are expected to be completed by the end of calendar year 2014.

Evaluation of Management's Response. We consider the Associate Administrator's proposed corrective actions responsive; therefore, the recommendation is resolved and will be closed upon completion and verification of the corrective actions.

Recommendation 3. We recommended that the Associate Administrator for the Human Exploration and Operations Mission Directorate routinely track adherence to the 90-day goal for responding to contractor requests for alternate requirement standards and variances and explore ways to facilitate the process in order to avoid delays or additional costs in development efforts.

Management's Response. The Associate Administrator concurred with our recommendation, stating that while the Program did not meet the 90-day goal for some of the Round 1 dispositions, in many cases this was to allow extra time for the partners to re-deliver products or provide additional supplementary information. Furthermore, according to the Associate Administrator, the Program was successful in dispositioning all deliverables by the end of September. To address our recommendation, the Associate Administrator stated that the Program completed a workshop in September 2013 to evaluate the lessons learned from Round 1 of the CPC Contract and to develop improvements to increase timeliness for Round 2. He also stated that NASA will begin tracking disposition timeliness against the 90-day goal and consider deferring the next round of reports to allow time for the the Agency to focus on key requests that may influence the partners' designs. The Associate Administrator anticipates completing these actions by the end of calendar year 2013.

Evaluation of Management's Response. We consider the Associate Administrator's proposed corrective actions responsive; therefore, the recommendation is resolved and will be closed upon completion and verification of the corrective actions.

Recommendation 4. We recommended that the Associate Administrator for the Human Exploration and Operations Mission Directorate, in conjunction with the FAA and the U.S. Air Force, formally establish a tri-agency Safety Steering Group for resolution of issues involving crew and public safety during commercial spaceflight operations in order to ensure that the Agency expeditiously meets all spaceflight safety issues.

Management's Response. The Associate Administrator concurred with our recommendation, noting that while NASA and the FAA have ongoing and comprehensive interactions regarding safety requirements and regulations, expanding this collaboration to include the Air Force in a more formal setting would be beneficial. The Associate Administrator stated that NASA has recently taken action to address our recommendation including obtaining approval from the Program Control Board to develop a Safety Steering Group that includes FAA and Air Force leadership, developing a draft charter for the Steering Group, and holding preliminary meetings between NASA, the FAA, and the Air Force's 45th Space Wing. The Associate Administrator also stated that additional meetings with the 45th Space Wing are scheduled for January 2014.

Evaluation of Management's Response. We consider the Associate Administrator's proposed corrective actions responsive; therefore, the recommendation is resolved and will be closed upon completion and verification of the corrective actions.

APPENDIX A

Scope and Methodology

We performed this audit from August 2012 through September 2013 in accordance with generally accepted government auditing standards. Those standards require that we plan and perform the audit to obtain sufficient, appropriate evidence to provide a reasonable basis for our findings and conclusions based on our audit objectives. We believe that the evidence obtained provides a reasonable basis for our findings and conclusions based on our audit objectives. This audit focused on the Commercial Crew Program managed by NASA's Commercial Spaceflight Development Office within the Human Exploration and Operations Mission Directorate. To determine the Agency's management of the Commercial Crew Program, progress made, and challenges hindering the successful implementation of the Program, we reviewed laws, regulations, and policies in order to determine compliance with required guidance and best practices. We obtained and reviewed prior reports and studies related to NASA's ability to address the development and collaboration challenges of the Commercial Crew Program. We interviewed key personnel within NASA's Commercial Spaceflight Development Office and the Commercial Crew Program located at NASA Headquarters, Kennedy Space Center, Johnson Space Center, and commercial partners at corporate sites or by telephone.

To determine whether the Commercial Crew Program is on track to provide a system capable of reaching the ISS in 2017, we analyzed the Program's documents, including budget documents, schedule timelines, and performance requirements as designed by NASA and presented to the development partners. We compared planned budgeting levels to actual budgeting levels and reviewed NASA's contingency planning for lower funding levels. We reviewed NASA's contingency plans to manage transportation to and from the ISS beyond 2016 in the case of major programmatic delays.

In order to assess the process used to communicate human rating requirements to the commercial partners, we reviewed human rated certification plans to assess how insight and oversight is to be implemented, and interviewed key NASA and commercial partner personnel. We also interviewed NASA Partner Integration Teams and observed their interactions with the commercial partners, and reviewed the requirements baseline to determine the variability of requirements modifications. Furthermore, we tracked and analyzed the timeliness of NASA's responses to the companies' requests for variances and alternate standards. By doing this, we were able to determine exactly how long it took NASA to provide a disposition on each of the requests, and therefore provide observations based upon the 90-day goal in the CPC contract, and from information provided by the companies.

Finally, to determine whether NASA is effectively coordinating with the FAA and other Government agencies to resolve spaceflight issues, we interviewed NASA personnel responsible for this coordination, applicable regulations and U.S. laws, as well as FAA personnel associated with certification and licensing of commercial spaceflight operations.

Use of Computer-Processed Data. We used computer-processed data to perform this audit. We collected computer-processed cost data related to the Commercial Crew Program in the form of milestone payment data from the beginning of the Program through the end of August 2013. Program officials downloaded the data from NASA's financial management program (Business Warehouse) and provided the data in Microsoft Excel. For our audit objectives, we compared this data to information provided in the President's budget estimates. We also queried the financial management program ourselves in order to validate the information received from the Program office. Therefore, we assess that the cost data we received is sufficiently reliable, but we did not rely solely on the computer-processed data to support our findings, conclusions, or recommendations.

Review of Internal Controls

We reviewed NASA policies and procedures to determine NASA's internal control requirements for managing the Commercial Crew Program. We found that while the Commercial Crew Program exhibited several good internal control practices, NASA officials should not have exempted this Program from adhering to required policies such as annual statements of assurance, periodic reviews by NASA's Independent Program Assessment Office, or formal assessments of their acquisition procedures as required by the Office of Management and Budget (OMB). We acknowledge that using Space Act Agreements requires some adjustment in procedures, but new programs and new approaches to development generally require more oversight, not less. Complying with basic OMB and GAO guidance helps managers provide reasonable assurance that government resources are safeguarded and errors or fraud are avoided. Although program officials understood the basic concept of internal control, the programs were not required by the Human Exploration and Operations Mission Directorate to perform internal control assessments. During the course of this audit, corrections were made and the Commercial Crew Program plans to submit a FY 2013 statement of assurance concerning their internal control program. In addition, although the Commercial Crew Program has spent a total of over $1 billion in government funding, the OMB-directed acquisition assessment for projects over $250 million has yet to be conducted. Finally, even though commercial spaceflight development commenced 7 years ago, neither the Cargo nor the Crew programs have been reviewed by NASA's Independent Program Assessment Office.[46] During the course of this audit, the Independent Program Assessment Office began designing a review of the Commercial Crew Program.

Prior Coverage

During the last 5 years, the NASA OIG and the GAO have issued 11 reports of particular relevance to the subject of this report. Unrestricted reports can be accessed over the Internet at http://oig.nasa.gov/audits/reports/FY13/index.html (NASA OIG) and http://www.gao.gov (GAO).

NASA Office of Inspector General

"Commercial Cargo: NASA's Management of Commercial Orbital Transportation Services and ISS Commercial Resupply Contracts" (IG-13-016, June 13, 2013)

"NASA's Challenges to Meeting Cost, Schedule, and Performance Goals" (IG-12-021, September 27, 2012)

"NASA's Challenges Certifying and Acquiring Commercial Crew Transportation Services" (IG-11-022, June 30, 2011)

"Review of NASA's Acquisition of Commercial Launch Services" (IG-11-12, February 17, 2011)

Government Accountability Office

"Commercial Space Launches: FAA Should Update How It Assesses Federal Liability Risk" (GAO-12-899, July 30, 2012)

"Commercial Space Transportation: Industry Trends, Government Challenges, and International Competitiveness Issues" (GAO-12-836T, June 20, 2012)

"NASA: Significant Challenges Remain for Access, Use, and Sustainment of the International Space Station" (GAO-12-587T, March 28, 2012)

"NASA: Assessment of Selected Large Scale Projects" (GAO-12-207SP, March 1, 2012)

"National Aeronautics and Space Administration: Acquisition Approach for Commercial Crew Transportation Includes Good Practices, but Faces Significant Challenges" (GAO-12-282, December 15, 2011)

"Key Controls NASA Employs to Guide Use and Management of Funded Space Act Agreements are Generally Sufficient, but Some Could Be Strengthened and Clarified" (GAO-12-230R, November 17, 2011)

"NASA: Commercial Partners Are Making Progress, but Face Aggressive Schedules to Demonstrate Critical Space Station Cargo Transport Capabilities" (GAO-09-618, June 16, 2009)

APPENDIX B

Commercial Partner Tables

Table 4. List of CPC Recipients

Company	System	Total Funds Available	Total FundsUsed (as of August 31, 2013)	Profile	Capability	Location
Sierra Nevada Corp	Dream Chaser	$10 million	$4million	Piloted lifting-body spacecraft	7 astronauts or equivalent crew and cargo; runway landing	Louisville, CO
The Boeing Company	CST-100	$9.9 million	$4million	Crew module and service module	7 astronauts or equivalent crew and cargo; dry surface landing	Houston, TX
Space Exploration Technologies (SpaceX)	Dragon	$9.6 million	$3.8 million	Free-flying, reusable spacecraft	7 astronauts; dry surface or ocean landing	Hawthorne, CA

Source: NASA OIG summary of Commercial Crew Program data

Table 5. List of CCiCap Recipients

Company	System	Total Funds Available[a]	Total Funds Used (as of August 31, 2013)	Profile	Capability	Milestones Completed (as of August 31, 2013)	Location
The Boeing Company	CST-100	$480 million	$280.9 million	Crew module and service module	7 astronauts or equivalent crew and cargo; dry surface landing	8 of 20	Houston, TX
Space Exploration Technologies Corporation (SpaceX)	Dragon	$460 million	$249 million	Free-flying, reusable spacecraft	7 astronauts; dry surface or ocean landing	7 of 16	Hawthorne, CA
Sierra Nevada Corporation	Dream Chaser	$227.5 million	$107.5million	Piloted lifting-body spacecraft	7 astronauts or equivalent crew and cargo; runway landing	4 of 12	Louisville, CO

[a] Total Funds Available includes optional milestones added in August 2013. Source: NASA OIG summary of Commercial Crew Program data.

Table 6. List of CCDev2 Recipients

Company	System	Total Funds Available	Total Funds Used (as of June 30, 2013)	Profile	Capability	Milestones Completed (as of June 30, 2013)	Location
The Boeing Company	CST-100	$112.9 million	$112.9 million (100%)	Crew module and service module	7 astronauts or equivalent crew and cargo; dry surface landing	16 of 16	Houston, TX
Sierra Nevada Corp	Dream Chaser	$105.6 million	$97.6 million (92%)	Piloted lifting-body spacecraft	7 astronauts or equivalent crew and cargo; runway landing	14 of 15	Louisville, CO
Space Exploration Technologies(SpaceX)	Dragon	$75 million	$75 million (100%)	Free-flying, reusable spacecraft	7 astronauts; dry surface or ocean landing	11 of 11	Hawthorne, CA
Blue Origin	Space Vehicle	$22 million	$22 million (100%)	Biconical spacecraft; Atlas V rocket initially, then reusable booster stage	4 astronauts	11 of 11a	Kent, WA

Table 6. (Continued)

Company	System	Total Funds Available	Total Funds Used (as of June 30, 2013)	Profile	Capability	Milestones Completed (as of June 30, 2013)	Location
Unfunded							
Alliant TechsystemsInc.	Liberty	n/a	n/a	Solid rocket boosters, Ariane 5 core stage, Vulcain 2 engine	44,500 lbs lift to low Earth orbit	5 of 5	Promontory, UT
Excalibur Almaz	Human Spacecraft	n/a	n/a	Human spacecraft, including a capsule, launch abort system, and expendable service module	3 astronauts or equivalent cargo	5 of 5	Houston, TX
United Launch Alliance	Atlas V	n/a	n/a	Atlas core stage, Centaur upper stage, and option for up to five solid rocket boosters	860,000 lbs of thrust from core stage at liftoff; ability to launch different crew spacecraft	5 of 5	Centennial,CO

[a] Three additional unfunded milestones were added in February 2013.
Source: NASA OIG summary of Commercial Crew Program data.

Table 7. List of CCDev1 Recipients

Company	System	Total Funds Spent	Profile	Capability	Milestones Completed (as of June 30, 2013)	Location
Sierra Nevada Corp	Dream Chaser	$20 million	Piloted lifting-body spacecraft	7 astronauts or equivalent crew and cargo; runway landing	4 of 4	Louisville, CO
The Boeing Company	CST-100	$18 million	Crew module and service module	7 astronauts or equivalent crew and cargo; dry surface landing	36 of 36	Houston, TX
United Launch Alliance	Atlas V	$6.7 million	Atlas core stage, Centaur upper stage, and option for up to five solid rocket boosters	860,000 pounds of thrust from core stage at liftoff; ability to launch different crew spacecraft	4 of 4	Centennial, CO
Blue Origin	Space Vehicle	$3.7 million	Biconical spacecraft; Atlas V rocket initially, then reusable booster stage	4 astronauts	7 of 7	Kent, WA
Paragon Space Development Corporation	Commercial Crew Transport-Air Revitalization System (CCT-ARS)	$1.44 million	n/a	Modular and reconfigurable life support system	5 of 5	Tucson, AZ

Source: NASA OIG summary of Commercial Crew Program data.

APPENDIX C

Comparison of Original and Revised Acquisition Strategies for the Commercial Crew Program

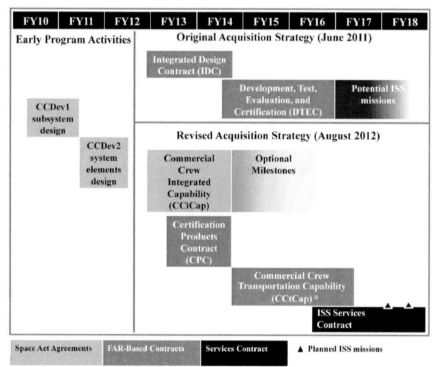

Figure 7. Comparison of Original and Revised Acquisition Strategies for the Crew Program.

Final report released by:
Paul K. Martin
Inspector General

Major Contributors to the Report:
Ridge Bowman, Director, Space Operations Directorate
Kevin Fagedes, Project Manager

Letisha Antone, Auditor
Janice Smith, Auditor
Michael Palinkas, Management Analyst
Troy Zigler, Management Analyst
Cedric Campbell, Associate Counsel to the Inspector General – Central Region

End Notes

[1] Space Act Agreements are a form of "Other Transaction Authority" provided to NASA in the National Aeronautics and Space Administration Act of 1958 that establish a set of legally enforceable commitments between NASA and a partner to accomplish a stated objective without imposing the extensive list of requirements routinely found in most government contracts.

[2] NASA OIG, "Commercial Cargo: NASA's Management of Commercial Orbital Transportation Services and ISS Commercial Resupply Contracts" (IG-13-016, June 13, 2013). In March 2013, SpaceX successfully completed its second Station resupply mission, while Orbital Sciences Corporation (Orbital) is scheduled to undertake its first resupply mission in December 2013.

[3] As defined by NASA, the Preliminary Design Review demonstrates that the overall preliminary design meets all requirements with acceptable risk and shows that the correct design options have been selected, interfaces have been identified, and verification methods have been described. The Critical Design Review determines if the integrated design is appropriately mature to continue to final design and fabrication. Both reviews are important to demonstrate that a system meets all requirements with acceptable risk and within cost and schedule constraints. For the Commercial Crew Program, the companies defined their own requirements for achieving both Preliminary and Critical Design Reviews, with review and concurrence by NASA. NASA funded Boeing and SpaceX to achieve Critical Design Review, but due to a limited budget has not funded Sierra Nevada's completion of that milestone.

[4] NASA has studies underway to determine if the lifespan of the ISS can be extended until 2028, but no firm decisions have been made regarding any extension.

[5] NASA, "NASA Cost Estimating Handbook" (2008).

[6] House Committee on Science, Space and Technology Hearing, "Recent Developments in NASA's Commercial Crew Acquisition Strategy" (September 14, 2012).

[7] NASA OIG, "Commercial Cargo: NASA's Management of Commercial Orbital Transportation Services and ISS Commercial Resupply Contracts" (IG-13-016, June 13, 2013). In March 2013, SpaceX successfully completed its second Station resupply mission, while Orbital is scheduled to undertake its first resupply mission in December 2013.

[8] NASA Policy Directive 1050.11, "Authority to Enter into Space Act Agreements" (December 23, 2008).

[9] Entry criteria are what a partner needs to perform or provide before reaching the milestone. Exit criteria are what must be performed in order to achieve successful completion of the milestone.

[10] Blue Origin completed the funded CCDev2 milestones and received final payment, but NASA and Blue Origin agreed to continue work until mid-2014 in an unfunded status.

[11] As defined by NASA, a Critical Design Review demonstrates that the maturity of the program's design is appropriate to support proceeding to full-scale fabrication, assembly, integration, and testing and that the technical effort is on track to complete the flight and ground system development and mission operations. For the Commercial Crew Program, the companies defined their own requirements for achieving Critical Design Review, with review and concurrence by NASA.

[12] On July 19, 2013, NASA released a draft Request for Proposal for the Phase 2 Certification under which it is proposing to use a FAR-based, firm-fixed-price contract for the final stages of development and certification. This includes a crewed system demonstration mission to the ISS and the potential to execute up to six crewed service missions. This phase is now called CCtCap.

[13] Although firm-fixed-price contracts are referenced in the Agency's acquisition strategy, Program managers said a final decision on the type of contract the Agency will use has not been made.

[14] For example, cost per flight in the U.S. Air Force's Evolved Expendable Launch Vehicle Program increased from $72 million in 1998 to $460 million in 2012. The Air Force originally awarded contracts to Boeing and Lockheed Martin, but in 2006, the companies created a joint venture known as the United Launch Alliance. In 2011, the Government Accountability Office (GAO) reported that a lack of competition coupled with a significant decrease in demand caused the increase in per-flight costs. In December 2012, the Air Force awarded SpaceX two Evolved Expendable Launch Vehicle-class missions and in June 2013 signed a cooperative agreement with the company to begin certifying an upgraded version of the Falcon 9 for the two missions – the first scheduled to take place in 2014 at a cost of $97 million and the second in 2015 at a cost of $165 million – less than half the cost of current United Launch Alliance missions.

[15] Specifically, the Space Act Agreements provide that "the goals of the CCiCap investments are to enable significant progress on maturing the design and development of an integrated commercial transportation system while ensuring crew and passenger safety."

[16] GAO, "NASA: Assessments Selected Large Scale Projects" (GAO-12-207SP, M arch 2012).

[17] A Preliminary Design Review demonstrates that the preliminary design meets all requirements with acceptable risk and shows that the correct design options have been selected, interfaces have been identified, and verification methods have been described.

[18] The CPC requires submission of four products: (1) alternate standards and variances (if applicable), (2) hazard reports, (3) verification and validation plans, and (4) certification plans. Each product was submitted to NASA in May 2013 for initial review, with final submissions due in February 2014.

[19] An integrated system review confirms that segments, components, and subsystems are available and ready to be integrated into the overall spaceflight system. In addition, such a review examines integration facilities, support personnel, and integration plans and procedures.

[20] SpaceX entered the commercial crew arena with a flight-tested rocket system. The company's Falcon 9 rocket and Dragon Capsule have flown four times to low Earth orbit and berthed to the ISS three times on cargo missions. Although not yet human rated, SpaceX designed the Dragon for astronaut transportation.

[21] Sierra Nevada is scheduled to complete Milestone 4 in late 2013, which will consist of one to five engineering test article free flight(s) to analyze aerodynamic data during the approach and landing phase of the Dream Chaser.

[22] By law, the ISS must be maintained until at least 2020, but NASA is examining the feasibility of extending the Station's life.

[23] At the end of March 2013, across-the-board budget reductions known as sequestration took effect, resulting in a 5 percent funding cut for this Program and most others across the Federal government. In addition, Congress imposed an additional 1.877 percent rescission against all Commerce, Justice, and Science agencies, which left the Program with $489 million in FY 2013. NASA subsequently submitted an operating plan in which it requested additional funds to address the shortfall and, as of August 2013, the operating plan was approved and the Program received $525 million – reprogrammed from other programs.

[24] As noted earlier in this report, Space Act Agreements are considered less costly because partners share costs.

[25] Science Subcommittee, U.S. House of Representatives Appropriations Committee, "NASA Oversight Hearing," YouTube (March 20, 2013), http://www.youtube.com/watch?v=5NEW1GIekYk (accessed August 20, 2013).

[26] GAO, "GAO Cost Estimating and Assessment Guide" (GAO-09-3SP, March 2009).

[27] For comparison, partner contributions for the cargo development program were roughly 50 percent.

[28] ASAP, "Annual Report for 2012" (January 9, 2013).

[29] Committee on NASA's Strategic Direction, Division on Engineering and Physical Sciences, National Research Council, "NASA's Strategic Direction and the Need for a National Consensus" (2012).

[30] NASA Procedural Requirements (NPR) 7120.5E, "NASA Space Flight Program and Project Management Requirements w/Changes 1-10" and NPR 7120.5 "NASA Space Flight Program and Project Management Handbook" (February 2010).

[31] An independent government cost estimate is conducted to check the reasonableness of a contractor's cost proposal and to make sure that the offered prices are within the budget range for a particular program. It documents the government's assessment of the program's most probable cost and ensures that enough funds are available to execute it. The Crew Program uses these types of estimates to inform budget planning efforts.

[32] House Committee on Science, Space and Technology Hearing, "Recent Developments in NASA's Commercial Crew Acquisition Strategy" (September 14, 2012). (Representative Donna F. Edwards and Vice Admiral Joseph Dyer-ASAP Chair).

[33] Booz Allen, "Independent Cost Assessment of the Commercial Crew Program: Independent Cost Assessment Report" (March 1, 2013).

[34] An integrated schedule is a time-based schedule containing the detailed tasks necessary to ensure successful program/contract execution. A risk register is a central repository for all risks identified by the project that includes information on probability, impact, and counter-measures.

[35] A partner may request an alternate standard that meets the intent of or is consistent with a NASA standard. Variances to requirements occur if a partner cannot meet a requirement, and are generally classified as exceptions, deviations, or waivers. NASA defines deviations as requests made during the formulation, planning, or design stages of a program to address expected situations and that provide temporary relief from a specific requirement in advance. NASA defines exceptions as permanent relief from a specific requirement, and they may be requested at any time during the life cycle of the program. NASA defines

  waivers as temporary relief from a specific requirement after the baseline system has been approved.

[36] NPR 8705.2B, "Human-Rating Requirements for Space Systems (w/change 4 dated 8/21/2012)."

[37] In addition to these documents, prior to December 2011, NASA also published the Commercial Orbital Transportation Services ISS Interface Requirements Document (SSP 50808), which outlines requirements for docking with the Station. NASA originally released SSP 50808 in September 2007 to provide international partners guidance on docking with the ISS.

[38] Phase I is CPC and Phase II is CCtCap.

[39] The supporting organizations include the Commercial Crew Program Systems Engineering and Integration Office; Program Control and Integration; Partner Integration; Mission Planning and Integration; Spacecraft Systems; Launch Vehicle Systems; Launch and Recovery Systems; Commercial Crew Program Chief Engineer; Commercial Crew Program Chief Safety Officer; Engineering; Safety and Mission Assurance; Commercial Crew Program Chief Health and Medical Officer; Space and Life Sciences; Flight Crew; ISS; Mission Operations – Johnson Space Center; Ground Processing – Kennedy Space Center; and the FAA (Advisory Role).

[40] The pressure suits would include everything needed for survival such as breathing oxygen, pressure exerted on the body, and even a heating and cooling system.

[41] 49 U.S. Code § 70101 et seq., Commercial Space Launch Act of 1984.

[42] 51 U.S. Code 509, paragraphs (a)(2) and (a)(5)(C) provided for the regulations to begin 8 years after the enactment of the Commercial Space Launch Amendments Act of 2004 (in 2012); this date was extended to begin October 1, 2015 by Public Law 112-95, February 14, 2012.

[43] NASA Policy Directive 8700.1, "NASA Policy for Safety and Mission Success" (October 28, 2008).

[44] NASA OIG, "NASA's Challenges Certifying and Acquiring Commercial Crew Transportation Services" (IG-11-022, June 30, 2011).

[45] The FAA and NASA must decide how claims for loss of life, injuries, or property damage will be handled, and/or the waivers of claims for missions that are contractor-operated and owned but involve NASA missions.

[46] On April 26, 2007, NASA's Program Management Council decided not to conduct an Independent Program Assessment Office review of the Commercial Cargo Program because of the uniqueness of the Program. The Cargo Program, now consisting of the Commercial Orbital Transportation Services and Commercial Resupply Services, has spent a total of over $1.7 billion in NASA funding.

In: NASA's Management of Commercial Crew ...   ISBN: 978-1-63484-065-1
Editor: Eileen Purcell   © 2016 Nova Science Publishers, Inc.

Chapter 2

COMMERCIAL CARGO: NASA'S MANAGEMENT OF COMMERCIAL ORBITAL TRANSPORTATION SERVICES AND ISS COMMERCIAL RESUPPLY CONTRACTS[*]

NASA Office of Inspector General

ACRONYMS

ATV	Automated Transfer Vehicle
CDR	Critical Design Review
COTS	Commercial Orbital Transportation Services
CRS	Commercial Resupply Services
FAR	Federal Acquisition Regulation
FY	Fiscal Year
GAO	Government Accountability Office
GPS	Global Positioning System
HTV	H-II Transfer Vehicle
ISS	International Space Station
LIDAR	Light Detection and Ranging

---

[*] This is an edited, reformatted and augmented version of a report, IG-13-016, issued by the National Aeronautics and Space Administration's Office of Inspector General, July 13, 2013.

| OIG | Office of Inspector General |
| PDR | Preliminary Design Review |

OVERVIEW

The Issue

In anticipation of the Space Shuttle's retirement, the NASA Authorization Act of 2005 directed NASA to develop cargo transportation capabilities to the International Space Station (ISS or Station) by fostering the commercial spaceflight industry. Reliable cargo transportation capabilities to the ISS are essential to ensure that critical life-sustaining supplies are provided to support the Station's crews and to maximize utilization of the ISS as a research lab by delivering and returning experiment-related materials to Earth. In the absence of commercial capabilities, NASA would need to rely on the spacecraft of international partners from Europe and Japan to resupply the ISS.

NASA's goals for its commercial cargo effort are to:

- implement U.S. Space Exploration policy with an investment to stimulate commercial enterprises in space;
- facilitate U.S. private industry demonstration of space cargo transportation capabilities with the goal of achieving reliable, cost effective access to low Earth orbit; and
- create a market environment in which commercial space transportation services are available to Government and private sector customers.

Given that the ISS is scheduled to be decommissioned in 2020, if NASA expects commercial partners to play a significant role in servicing the ISS, timely and successful development of their transportation capabilities is crucial.[1]

To foster commercial cargo capabilities and procure ISS resupply missions, NASA used a combination of Space Act Agreements as part of its Commercial Orbital Transportation Services (COTS) Program and fixed-price contracts structured in accordance with the Federal Acquisition Regulation (FAR) as part of its Commercial Resupply Services (CRS) Contract.[2] For the COTS Program, NASA collaborated with and provided funding to two private companies – Space Exploration Technologies Corporation (SpaceX) and

Orbital Sciences Corporation (Orbital) – to further the companies' development of spaceflight cargo capabilities.[3] SpaceX and Orbital have shared costs with NASA by contributing more than 50 percent of the funds needed for spacecraft development.

In 2008, NASA awarded $3.5 billion in fixed-price contracts to SpaceX ($1.6 billion for 12 missions) and Orbital ($1.9 billion for 8 missions) to procure resupply services to the ISS.[4] In October 2012 and March 2013, SpaceX successfully completed two resupply missions. Orbital plans to undertake its first resupply mission in late 2013. NASA officials stress that the success of both SpaceX and Orbital is critical to the sustainability and utilization of the ISS.

NASA's commercial cargo program is at a critical stage with Orbital poised to come online later this year and the scheduled decommissioning of the ISS in 2020 rapidly approaching. Meanwhile, NASA is using a similar acquisition strategy – a combination of Space Act Agreements and FAR-based, fixed-price contracts – to pursue commercial transportation services to the ISS for its astronauts.[5] To this end, the successes and challenges experienced by NASA's commercial cargo program will prove to be instructive to its commercial crew effort.

Given the importance of the commercial cargo program to the continued viability of the ISS, we examined NASA's management of the program. Details of the audit's scope and methodology are in Appendix A.

Results

Following a nearly 3-year delay early in development, SpaceX successfully completed its final system demonstration flight and two resupply missions to the ISS. Although each flight experienced some anomalies, none were serious enough to substantially impact the missions. During the final demonstration flight, SpaceX needed to adjust its Dragon capsule's guidance system, causing a short delay in the capsule's final approach to the ISS. During its first cargo mission, SpaceX encountered a failure on one of the nine Merlin engines in its Falcon 9 rocket, several hardware failures caused by radiation exposure, three instances of sensors losing functionality in the Dragon's thrusters, and the loss of all three coolant pumps due to a water leak after splashdown in the ocean. All radiation effects were resolved with no mission impact, the faulty temperature sensors represented a loss of redundancy only, and the failure of the coolant pumps did not affect the science experiments on board in the return payload. However, these issues contributed to a 2-month delay of the launch of the second cargo mission, which was moved from January to March 2013.

During the second cargo mission, a malfunction initially limited the operation of three of the four thruster pods used to boost the Dragon to a higher orbit and perform the final maneuvers necessary to rendezvous with the ISS. The problem was corrected and the Dragon berthed with the ISS one day later than scheduled with no operational impact to the mission. As of the end of fiscal year (FY) 2012, NASA had paid SpaceX $858 million for development and cargo resupply services under both its funded Space Act Agreement and FAR-based contract.

Like SpaceX, Orbital has experienced delays of over 2 years in its COTS Program, including an early change from an unpressurized to pressurized capsule and construction delays on its Wallops Island, Virginia, launch facility. NASA has paid Orbital a total of $910 million as of the end of FY 2012, including funding for both development efforts under its COTS Space Act Agreement and CRS contract. Under the current payment schedule, the company is on track to receive up to 70 percent of the funds associated with six of its eight CRS missions prior to having flown a demonstration flight.[6]

Orbital successfully completed a maiden test flight of its Antares rocket on April 21, 2013, but the full demonstration flight required under the COTS Program most recently scheduled for June 2013 has slipped to August or September 2013. NASA and Orbital officials noted the maiden flight has reduced technical risk and that the costs of any system modifications needed as a result of the demonstration flight will be borne by Orbital given that the CRS contract is fixed price. Nevertheless, the possibility remains that the demonstration flight could expose issues that require costly rework and redesign, resulting in major adjustments to the current CRS launch schedule.

Out of a need to ensure a redundant cargo capacity, NASA funded development of SpaceX's and Orbital's spaceflight capabilities under the COTS Program while concurrently funding fabrication of the companies' spacecraft under the CRS contracts. As a general matter, procuring rocket systems prior to a successful system demonstration flight substantially increases financial risk as major technical problems may be encountered during final testing and demonstration. Although CRS activities are not contractually tied to a successful demonstration flight, as Orbital's COTS development activities slipped so did the anticipated launch dates for its CRS missions. Although we do not second guess NASA's decision to concurrently fund up to three rocket systems given the critical need for additional ISS resupply capabilities, in the case of Orbital, NASA will fully or partially fund six rocket systems under the CRS contract before Orbital has fully demonstrated its spaceflight system.[7]

In our judgment, NASA has been too slow to adjust its payment schedule to Orbital under the CRS contract given the substantial slippage in the launch schedule for the company's resupply missions. As such, given the risks inherent in concurrent development, we question NASA's decision to pay Orbital approximately $150 million for costs associated with their fourth and fifth resupply missions. We believe NASA should have deferred this amount to future fiscal years in order to avoid spending funds too far in advance of each mission's launch dates. During the course of our review, NASA took steps to adjust its payment schedule in light of the development delays by negotiating a contract modification in December 2012 for Mission 6 that tied payment to a successful Antares maiden test flight. In our view, NASA instead should have tied payment for this mission to a successful full system demonstration flight. Finally, Orbital requested to begin work on resupply Mission 7 by May 2013, a request from our perspective that, if approved, would result in an additional estimated $70 million in premature payments to the company in FY 2013.[8]

We discussed with program officials our concerns about these advance procurements and they recognized the need to slow down the pace of NASA's payment for Orbital's rocket systems production. For example, officials said they tied future Mission 4 and 5 payments to an adjusted launch schedule and completion of ISS integration activities by the company.[9] While we appreciate that NASA has taken these steps, we believe the Agency has accepted too much financial risk by funding Orbital's fabrication of rocket systems for Missions 4, 5, and 6 so far in advance of the time needed to meet the ISS resupply schedule and prior to Orbital completing a successful system demonstration flight.

Management Action

To reduce the Agency's financial risk, we recommended that the Associate Administrator for Human Exploration and Operations Mission Directorate ensure that contractual agreements for the commercial cargo providers are updated to reflect the lead times required to meet any revised launch dates. If launch dates slip, NASA should adjust contract work plans to ensure that the authorized lead times and NASA payments reflect the revised schedules. In response to our draft report, the Associate Administrator concurred with our recommendation. We consider the Associate Administrator's proposed actions to be responsive to our recommendation and will close the recommendation upon completion and verification of the corrective actions.

While the Associate Administrator concurred with the recommendation, he disagreed that NASA has accepted too much financial risk in the way it has implemented the Orbital CRS contract. He stated that NASA determined that the programmatic risks of not starting hardware development needed for cargo resupply were substantially greater than the financial risks posed to the Agency by doing so. He further stated that NASA uses existing payment cap protections and other contractual provisions to reduce financial risks and align payments with technical performance.

We agree that balancing programmatic and financial risk is critical to ensure the success of the commercial cargo program. However, as outlined in the report we continue to believe that NASA has been too slow to adjust its payment schedule to Orbital given the substantial slippage in the launch schedule for the company's resupply missions.

INTRODUCTION

Background

In 2004, President George W. Bush announced the Vision for Space Exploration, which, among other initiatives, directed NASA to pursue access to the International Space Station (ISS) and low Earth orbit for both crew and cargo by means of commercial partners. Congress responded by enacting the NASA Authorization Act of 2005, which directed the Agency to facilitate agreements with U.S. companies for the research and development of commercial spaceflight capabilities. That same year, NASA created the Commercial Crew and Cargo Program Office to stimulate efforts within the private sector to develop safe, reliable, and cost-effective transportation capabilities.

Between 2006 and 2012, NASA spent $743 million on commercial cargo development efforts through its Commercial Orbital Transportation Services (COTS) Program. In addition, since 2009 the Agency has made $1.1 billion in payments in connection with two fixed-price contracts for cargo transportation services to and from the ISS through its Commercial Resupply Services (CRS)

Contract. Two companies – Space Exploration Technologies Corporation (SpaceX) and Orbital Sciences Corporation (Orbital) – received development funds under the COTS Program and are also executing task orders and work plans under the CRS contracts.[10]

Space Act Agreements. The National Aeronautics and Space Act of 1958 (as amended) gives NASA authority to utilize agreements to work with nongovernment entities other than the standard contracts governed by the Federal Acquisition Regulation (FAR). One of the most frequently used of these "Other Transaction Authorities" is Space Act Agreements, which establish a set of legally enforceable commitments between NASA and a partner but do not incorporate an extensive list of requirements routinely found in FAR contracts.[11] In Space Act Agreements, NASA agrees to provide funding, goods, services, facilities, or equipment that the partner uses to accomplish stated objectives. In return, the partner may advance technologies that support NASA's mission, share information, or reimburse NASA for the support provided. As of March 2013, NASA reported over 1,500 active Space Act Agreements with various private companies, educational institutions, state and Federal government organizations, and foreign governments and entities.

The Space Act provides authority for reimbursable, nonreimbursable (unfunded), and funded agreements, and NASA has utilized all three types of agreements in its commercial cargo development program. In a funded agreement, NASA transfers appropriated funds to a partner to accomplish a specific objective. The Agency may use funded agreements only if it cannot achieve its objectives through other types of instruments, including traditional FAR-based contracts. In nonreimbursable agreements, each party bears the cost of its participation with no exchange of funds, while in reimbursable agreements a partner reimburses NASA for support provided such as test facilities, supplies, or engineering expertise.[12]

Commercial Orbital Transportation Services. NASA's commercial cargo development program – known as COTS – aims to stimulate efforts within the private sector to develop and demonstrate safe, reliable, and cost-effective cargo transportation capabilities to the ISS and low Earth orbit. In 2006, NASA announced it would begin this effort by competitively funding multiple Space Act Agreements. That year, NASA entered into Space Act Agreements with SpaceX and Rocketplane Kistler for $278 million and $207 million, respectively. However, because Rocketplane Kistler failed to achieve several financial and technical milestones, NASA terminated its Agreement with the company in 2007 after expending $32.1 million. In February 2008, NASA recompeted the opportunity and entered into a Space Act Agreement with Orbital worth up to $170 million. In fiscal year (FY) 2011, NASA added additional milestones to the SpaceX and Orbital Agreements, bringing the total value of the respective Agreements to $396 million and $288 million.

Both SpaceX and Orbital were expected to contribute their own funds to the commercial cargo development efforts and to date have contributed more than 50 percent of overall development costs. In essence, this means that NASA and the companies share financial, schedule, and technical risks associated with the Program. In addition, both agreements called for the companies to demonstrate a cargo resupply capability in less than 3 years – a substantially shorter timeframe than the average 5 to 6 years NASA officials told us such efforts typically take to complete.

As a part of the Space Act Agreement negotiations, NASA and the companies agreed to a series of developmental milestones that tied payments to events such as design reviews, testing, and ultimately full system demonstration flights. Once NASA and the company agree that a particular milestone has been accomplished, NASA pays the company a predetermined amount associated with that milestone.

A key milestone in the Space Act Agreements for both companies is a full system demonstration flight, which includes successful launch and rendezvous with the ISS. For SpaceX, the full system demonstration flight also included successful delivery of cargo to and the return of cargo from the ISS.[13] Initially, SpaceX proposed to conduct three demonstration flights of its rocket systems under the COTS Agreement. Ultimately, SpaceX self-funded the maiden flight of its Falcon 9-Dragon transportation system. That mission was followed by two demonstration missions under the COTS Agreement, one of which included berthing with the ISS. Given the success of these demonstration flights, NASA determined that SpaceX had met or exceeded all of the technical objectives in the COTS Agreement and thus did not require the company to perform a third demonstration mission.

Orbital's Space Act Agreement includes a single system demonstration flight currently planned for August or September 2013. In 2011, NASA added a milestone to the Agreement linked to a maiden test flight of Orbital's Antares rocket. Orbital successfully completed this test flight on April 21, 2013.

Commercial Resupply Services Contracts. In December 2008, while SpaceX's and Orbital's development efforts were still underway pursuant to their COTS Space Act Agreements, NASA entered into a separate set of FAR-based, firm-fixed-price contracts with the companies for a series of resupply missions to the ISS. Worth a total of $3.5 billion, these contracts are for 20 cargo resupply missions through at least 2016: 12 by SpaceX and 8 by Orbital. (See Appendix B for more information concerning the CRS contracts.) Under the contracts, NASA and the companies agreed to a price per mission for the delivery to and return of cargo and disposable items from the ISS. Because costs are expected to rise over the life of the program, annual increases to the price per mission are factored into the contracts.

NASA uses task orders and work plans to manage the CRS contracts. Task orders are issued to the companies for specific projects related to a mission. To date, NASA has issued 26 task orders: 18 to SpaceX and 8 to Orbital.[14] For both companies, Task Order 1 describes the basic requirements for the 20 resupply missions. Work plans are published for each mission and identify the milestone dates tied to specific payments by NASA to the companies. Each work plan is designed to allow sufficient lead time to build the vehicle to meet the ISS resupply schedule. Because NASA wanted to start resupply missions as soon as possible, NASA initiated its FAR-based resupply contracts with the understanding that the companies would receive authority and funds to build rocket systems prior to completing their first demonstration flights. Due to the imminent need for cargo missions and the lead time necessary to fabricate the systems, NASA accepted the risk associated with building vehicles before a successful demonstration flight. The partners sought to help mitigate this risk by testing individual systems and components prior to the demonstration flight.

When SpaceX received its CRS contract award in 2008, its COTS efforts had been underway for 28 months and the company had completed its system design reviews. In contrast, NASA's Space Act Agreement with Orbital had been in place for just 10 months and its system designs were still preliminary when Orbital was awarded its CRS contract in 2008. Moreover, once the requirements of the CRS contract were finalized, Orbital altered its COTS demonstration mission design for an unpressurized capsule to address NASA's request for a pressurized capsule.[15]

The CRS contracts were not directly tied to the milestones outlined in the companies' COTS Agreements. Rather, they were tied to a launch schedule that anticipated the companies' development efforts would be completed with full demonstration flights approximately 2-3 years later. However, as developmental efforts began to slip and completion of COTS milestones was delayed, this assumption turned out to be incorrect. Moreover, although NASA has authority under the CRS contract to adjust the payment schedule based on revised launch dates, we found this was not routinely done in a timely manner.

SpaceX launches from Cape Canaveral Air Force Station in Florida, adjacent to NASA's Kennedy Space Center. Following a series of delays, SpaceX flew a successful maiden flight of its Falcon 9-Dragon system in June 2010, a full system demonstration of its launch system in December 2010, a demonstration flight to the ISS in May 2012, and two resupply missions to the ISS in October 2012 and March 2013.

	Orbital Sciences Corporation	Space Exploration Technologies Corporation

Launch Vehicle/Capsule	Antares/Cygnus	Falcon 9/Dragon
Launch Site	Wallops Flight Facility, VA	Canaveral AFS, FL
Operational Status	Maiden test flight: APR 13; Demonstration: AUG/SEP 13	Maiden test flight: JUNE 10; Demonstrations: DEC 10 and MAY 12; 2 of 12 CRS missions to ISS completed
Cargo Capability To ISS From ISS	1,700-2,700 Kilograms 2,000 kilograms (Disposed)[a]	3,310 Kilograms 3,310 Kilograms[b]

[a] Orbital's Cygnus capsule will have the capacity to carry 1,700 kilograms in its first three resupply missions and 2,700 kilograms on the remaining five missions. In addition, up to 2,000 kilograms of trash from the ISS can be disposed of by loading on the capsule, which will burn up upon reentry.
[b] SpaceX can return up to 2,500 kg of cargo to Earth, or a total of 3,310 kg of pressurized (return) and unpressurized (disposal) cargo on a single mission.

Source: NASA Office of Inspector General (OIG) presentation of company and Program information.

Figure 1. Commercial Orbital Transportation Services – Orbital and SpaceX.

Orbital's rocket system launches from a newly constructed pad at Wallops Flight Facility in Virginia. Orbital conducted a successful hot fire test of its Antares engines in February 2013 and a test flight of its Antares rocket in April 2013.

A demonstration mission of its full system to the ISS is planned for August or September 2013. Orbital's first resupply mission to the ISS is scheduled for the last quarter of calendar year 2013. Figure 1 provides a summary of both companies' spaceflight systems.

ISS Cargo Transportation Providers. In addition to the SpaceX and Orbital vehicles, NASA has access to two other vehicles to transport supplies to the ISS: the European Space Agency Automated Transfer Vehicle (ATV) and the Japanese Aerospace Exploration Agency H-II Transfer Vehicle (HTV).[16]

The international partners and NASA worked together to produce a schedule that meets ISS needs and maximizes each vehicle's capabilities. NASA officials stated that because European cargo missions are scheduled to end in mid-2014, these vehicles would not provide sufficient capability to meet the Agency's ISS cargo transportation needs beginning in 2015. In addition, Program officials said the last two scheduled HTV flights in 2015 and 2016 are slated to carry 24 primary batteries for the ISS, which significantly reduces the available pressurized resupply capability of these flights.

SpaceX is the sole transportation provider capable of returning equipment, supplies, and research experiments to Earth, as all the other vehicles (including Orbital's Cygnus) are designed to disintegrate upon reentering the Earth's atmosphere. Figure 2 provides the projected schedule for cargo resupply flights to the ISS through 2016. Missions beyond 2016 are not yet scheduled.

NASA barters with the European Space Agency and the Japanese Aerospace Exploration Agency for cargo transportation on their vehicles.[17] As of January 2013, the Europeans have flown three of five planned cargo missions on the current schedule, while the Japanese have flown three of seven planned missions.

The costs per mission for cargo resupply both by SpaceX and by Orbital are expected to be lower than the costs associated with the European and Japanese vehicles. Table 1 compares the average cost per mission and capabilities of all ISS cargo service providers.

	2012	2013	2014	2015	2016
Orbital Cargo Missions (Cygnus)		D 1 2 3 / Aug Sep Nov Jan	4 5 / May Oct Feb	6 / Sep	7 8 / Mar Sep
SpaceX Cargo Missions (Dragon)	D 1 / May Oct	2 3 / Mar Nov	4 5 6 7 8 9 / Apr Aug Nov Feb Apr Aug		10 11 12 / Jan Apr Aug
European Cargo Missions (ATV)	3 / Mar	4 / June	5 / Apr		
Japanese Cargo Missions (HTV)	3 / Jul	4 / Aug	5 / Jul	6 / Jul	7 / Jul
Russian Cargo Missions (Progress)*	46 47 48 49 / Jan Apr Aug Oct	50 51 52 53 / Feb Apr Jul Oct	54 55 56 57 / Feb Apr Jul Oct	58 59 60 61 / Feb Apr Jul Oct	62 63 64 65 / Feb Apr Jul Oct

*At this time, the United States does not procure cargo resupply services from Russia.

ATV – Automated Transfer Vehicle   Numbers represent launch number.
HTV – H-II Transfer Vehicle   D is for demonstration launch.

Source: NASA OIG presentation of Program information.

Figure 2. ISS Cargo Mission Schedule as of May 2013.

Table 1. Cargo Transportation Average Price per Mission and Capability

	Contract award (billions)	Number of missions on contract	Average price per mission (millions)	Potential pressurized and/or unpressurized payload delivered to ISS per mission	Payload returned from ISS or disposed per mission
Orbital	$1.9	8	$237.5	1,700 kg standard[a] 2,700 kg enhanced	2,000 kg (disposed[b])
SpaceX	$1.6	12	$133.3	3,310 kg	2,500 kg pressurized (returned[c]) 3,310 kg (pressurized + unpressurized[d])
International Partner Capabilities[e]					
JAXA – HTV	-	-	$400 - 450	6,000 kg max [5,200 kg (pressurized) 1,500 kg (unpressurized)]	< 6,000 kg (disposed)
ESA – ATV	-	-	$450 - 500	7,667 kg	6,500 kg (disposed)

[a] Orbital's first three resupply missions will use the standard Cygnus cargo module; the rest will use the enhanced module, which can hold an additional 1,000 kilograms (kg) of cargo.
[b] Orbital's return capability consists of waste disposal only.
[c] SpaceX has the ability to return cargo such as science experiments to Earth intact.
[d] SpaceX also has the ability to return a combination of pressurized and unpressurized cargo, part of which returns to Earth intact, and the rest is disposed.
[e] Average price per mission for JAXA and ESA was taken from open source publications, and is not the price NASA pays due to barter agreements.

Source: NASA OIG analysis of Program information.

Objectives

Given the importance of NASA's commercial cargo program to support the ISS through the remainder of its scheduled lifespan, we assessed the program's progress to date and the Agency's overall management approach. Specifically, we assessed the extent to which the Agency and its commercial partners are on track to resupply the ISS. See Appendix A for details of the audit's scope and methodology, our review of internal controls, and a list of prior audit coverage.

RESULTS

Spacex Successfully Completed Its Demonstration Flights and Two Resupply Missions to the ISS

Following a nearly 3-year delay in development, SpaceX successfully completed its final system demonstration flight and two resupply missions to the ISS as of April 2013. Although each flight experienced technical anomalies, none were serious enough to substantially impact the missions and, according to ISS Program officials, were fewer in number and complexity than what program managers encountered during other space programs. For example, during the second demonstration flight SpaceX needed to adjust Dragon's guidance system prior to its final approach to the ISS. During the company's first cargo mission, SpaceX experienced a failure in one of its nine engines, several hardware failures in the Dragon caused by radiation exposure, three instances of sensors losing functionality in the Dragon's thrusters, and the loss of all three coolant pumps due to a water leak after splashdown in the ocean. All radiation effects were resolved with no mission impact, the faulty temperature sensors represented a loss of redundancy only, and failure of the coolant pumps did not lead to loss of science experiments on the return payload. However, these issues contributed to a 2-month delay for the second cargo mission, which slipped from January to March 2013. During the second cargo mission, a malfunction initially limited operation of three of the four thruster pods used to boost the Dragon to a higher orbit and perform the final maneuvers necessary to rendezvous with the ISS. The problem was quickly corrected, and the Dragon berthed with the ISS one day later than scheduled with no operational impact.

SpaceX Successfully Demonstrated its System Albeit Nearly 3 Years Later than Originally Scheduled

NASA awarded SpaceX a $278 million Space Act Agreement as part of the COTS Program in August 2006, and 2 years later a $1.6 billion firm-fixed-price CRS contract for 12 resupply flights to the ISS. In FY 2011, NASA added milestones to the Space Act Agreement, bringing its total value to $396 million. With its May 2012 demonstration flight, SpaceX satisfied all requirements of the Space Act Agreement and received its final milestone payment in August 2012.

Launch contracts are typically paid in increments tied to the successful completion of production milestones. As of the end of FY 2012, NASA had paid SpaceX $462 million on its CRS contract (see Table 2). This included full payment for the company's first ISS resupply mission and partial funding for completed milestones associated with the next five missions. Work on a seventh mission began in December 2012.

Table 2. Summary of COTS and CRS Funding for SpaceX

	Agreement type	Total award (millions)	Total spent ᵃ (millions)	Percent spent
Commercial Orbital Transportation Services (COTS)	Space Act Agreement	$396	$396	100%
Commercial Resupply Services (CRS)	Fixed-price contract	$1,600	$462	29%
Total		$1,996	$858	43%

ᵃ Dollars as of the end of FY 2012.
Source: NASA OIG analysis of Program information.

Originally, SpaceX was scheduled to conduct three demonstration missions under its Space Act Agreement, with the final mission scheduled for September 2009.[18] However, technical issues – including those related to design and software development – caused SpaceX to delay its first COTS demonstration flight until December 2010. Given the success of this flight and the second COTS demonstration, NASA did not require a third COTS demonstration, thereby finalizing the company's obligations under its COTS Agreement. The delay in the final demonstration mission shifted the schedule for the company's first resupply mission from mid-2010 to October 2012 (see Figure 3). The company's second mission successfully launched on March 1, 2013, berthed to the ISS with the assistance of the Station's robotic arm, and returned to Earth on March 26, 2013.

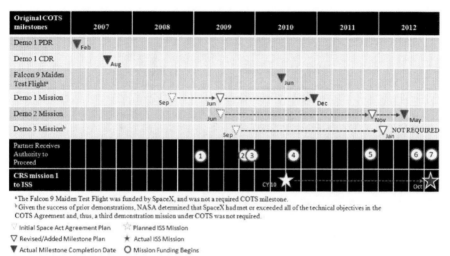

Source: NASA OIG analysis of Program information.

Figure 3. Key Milestones and Test Event Delays for SpaceX.

COTS Demonstration Flight Was Successful

SpaceX's May 2012 flight successfully demonstrated the company's ability to launch the Dragon capsule (see Figure 4) and to approach and berth with the ISS. During the flight, SpaceX experienced a minor technical issue with the Dragon's Light Detection and Ranging (LIDAR) guidance system used in conjunction with a thermal imaging sensor to guide the capsule toward the ISS. As the Dragon capsule made its way to the ISS, SpaceX halted its progress due to unanticipated LIDAR reflections from the ISS structure that were removed by reducing the field of view of the sensor. This action allowed the Dragon capsule to properly analyze the distance from the ISS, safely approach the Station, and attach to a docking port with the help of the Station's robotic grappling arm. The LIDAR guidance system functioned correctly during SpaceX's October 2012 resupply flight.

Technical Issues During SpaceX Mission 1

In October 2012, SpaceX launched CRS Mission 1, the first of its 12 cargo resupply flights to the ISS. Mission 1 carried approximately 450 kilograms of supplies, including 166 scientific investigations.[19] In addition, the mission carried a secondary payload – a communications satellite for the company ORBCOMM. As is common, the flight experienced several minor technical issues, none of which affected the success of the NASA cargo mission.

Source: NASA.

Figure 4. SpaceX's Dragon Capsule.

First Stage Engine Failure. SpaceX's Falcon 9 first stage is powered by nine Merlin engines. Approximately 79 seconds into launch, Falcon 9 experienced a failure in one engine, which automatically shut down. The Falcon 9 is designed to operate in the event of the failure of one engine, and the Dragon capsule remained on course to its rendezvous with the ISS. However, the engine failure resulted in the loss of the prototype ORBCOMM communications satellite, which deployed at a lower than expected altitude and was unable to reach its operational orbit.[20] SpaceX subsequently conducted an investigation and NASA received a report on the failure in February 2013.

For the second CRS mission, SpaceX used engines that had been thoroughly inspected to preclude recurrence of the problem. Moreover, SpaceX officials stated that the upgraded engine that will be used in future flights does not use the same production process.

Flight Computer Malfunction and Radiation Events. In order to decrease computing power, facilitate ability to upgrade, and increase processor speed, SpaceX decided not to use radiation-hardened parts, and instead chose to address radiation effects through a radiation-tolerant avionics system and design recovery strategies if a failure were to occur.[21] For example, to compensate for the danger of a radiation-induced malfunction, SpaceX uses three redundant computers to ensure that one or more remain operational.

Source: NASA.

Figure 5. Dragon Splashdown to End Cargo Mission 1, October 28, 2012.

Following its successful berthing to the ISS in October 2012, the Dragon experienced a number of radiation-related malfunctions. First, following a wave of solar radiation, one of the capsule's three computers desynchronized from the other two. After the malfunctioning computer successfully rebooted, SpaceX technicians decided not to resynchronize it with the other computers operating the capsule. Because the capsule was able to operate at full capability with only two computers – the minimum required by NASA – Agency officials said that re-synchronizing the failed computer was not worth the risk, and the mission proceeded using the two computer systems. NASA and SpaceX subsequently developed contingency plans to address this situation were it to occur on SpaceX Missions 2 or 3. The Dragon capsule that SpaceX is preparing for Mission 4 will be able to resynchronize its computers without rebooting.

In addition to the malfunction of the flight computer, radiation also caused the temporary loss of one of three Global Positioning System (GPS) sensors, the propulsion and trunk computers, and an ethernet switch. Once the Dragon reached orbit and prior to berthing with the ISS, the GPS sensor suffered a radiation hit and lost power. However, technicians were able to quickly restore power and restart the sensor. While berthed to the ISS, the propulsion and trunk computers as well as an ethernet switch lost power due to radiation exposure. Both systems fully recovered following a restart.

Capsule Thruster Sensor Malfunctions. The Dragon capsule uses thrusters to control attitude and maneuvering capabilities once it has entered low Earth orbit. During CRS Mission 1, sensors on two thrusters malfunctioned. On Thruster 3, a sensor reading drifted for a short period. SpaceX technicians monitored the thruster for the remainder of the flight and did not report any additional or undesired drift. On Thruster 4, a temperature sensor reading also had a similar error.

NASA officials said that each of the sensors had backups, which alleviated any maneuvering issues. In addition, the malfunctioning sensors did not fall below NASA-required performance thresholds, allowing the thrusters to continue normal operations.

Electrical Power Lost After Splashdown. The Dragon capsule returned from CRS Mission 1 on October 28, 2012, with 393 kilograms of scientific experiments and 235 kilograms of hardware. After splashdown in the Pacific Ocean (see Figure 5) and before the Dragon capsule could be lifted onto the recovery ship, water leaked into external portions of the capsule, which prompted SpaceX officials to turn off electrical power to several components, avoiding potential damage. As a result, three coolant pumps that maintain temperatures inside the cabin were shut off, although temperatures remained within required limits. In addition, electrical power to the General Laboratory Active Cryogenic ISS Experiment Refrigerator (GLACIER) refrigeration system was turned off. The refrigerator was storing temperature-sensitive samples from investigations conducted on the ISS. Although temperatures inside the refrigerator rose 30 degrees from the desired temperature of minus 95 degrees Celsius, none of the materials stored inside this device were damaged.

NASA officials explained that the water seepage was caused by the loss of air pressure in portions of the capsule during deorbit, creating a vacuum at splashdown. The capsule used during the company's May 2012 demonstration flight also experienced seepage after splashdown and SpaceX began to take steps to address the problem following that flight. However, by the time of the demonstration flight, the vehicles for use in CRS Missions 1 and 2 were nearly complete and little could be done to modify them. Accordingly, SpaceX modified the storage containers for the second flight in an attempt to increase their resistance to water seepage. SpaceX has since modified the design of the capsule for the third mission to guard against water intrusion.

As a result of the technical issues experienced on CRS Mission 1, SpaceX and NASA rescheduled the second resupply mission from January to March 2013. None of these issues turned out to be major problems, and NASA officials did not consider the 2-month delay substantial or indicative of future negative performance issues.

Technical Issue Briefly Delayed Berthing with ISS during SpaceX Mission 2

On March 1, 2013, SpaceX launched the second of its 12 CRS missions to the ISS with 677 kilograms of supplies, including 160 scientific experiments. At approximately 9 minutes into launch, a minor issue occurred with the Dragon capsule's oxidizer tank pressure. This malfunction limited the operation of three of the four thruster pods used to boost the capsule to a higher orbit and perform the final maneuvers necessary to rendezvous with the ISS. SpaceX delayed deployment of the Dragon's solar array for 90 minutes until technicians determined that deployment would not exacerbate the thruster problem. NASA has a firm requirement that at least three thruster pods be operational before the Dragon can approach the ISS. Accordingly, SpaceX delayed further ascent until it corrected the thruster problem. Approximately 5 hours after launch, SpaceX engineers corrected the issue with the oxidizer tanks and their associated plumbing, thereby returning all four thruster pods to normal operation. The Dragon recomputed its ascent and rendezvous profiles and resumed its approach to the ISS.

The Dragon successfully berthed to the ISS on March 3 – one day later than scheduled. This delay did not adversely affect any experiments aboard the Dragon capsule. The capsule remained at the ISS until March 26, when it returned to Earth with 1,370 kilograms of science samples.

NASA Paid Orbital Significant Funding under the Resupply Contract for Six Missions Even though the Company Has Yet to Complete Its First Flight to the ISS

Like SpaceX, Orbital has experienced delays in its COTS development program, including postponement of critical system reviews and flight tests. Despite these delays, NASA has paid Orbital a total of $910 million as of the end of FY 2012, including funding for both COTS development efforts and cargo resupply services under its CRS contract. As of May 2013, the company is on track to receive up to 70 percent of the funds associated with six of its eight CRS resupply missions prior to having flown a demonstration flight.[22] Given the risks inherent in concurrent development and production, we question NASA's decision to pay Orbital approximately $150 million for costs associated with Orbital's fourth and fifth resupply missions and believe NASA paid these funds too far in advance of the missions' launch dates. These actions increased NASA's financial risk in the event that the system demonstration flight reveals the need for design changes and modifications to Orbital's rocket system. Moreover, in December 2012 NASA gave Orbital authority to proceed on Mission 6 with payments contingent upon a successful maiden test flight of the company's Antares rocket. In our view, NASA instead should have made payment for this mission contingent upon a successful full system demonstration flight. Finally, as part of their Launch on Need proposal, Orbital requested to begin work on resupply Mission 7 by May 2013, a request that if approved would result in an additional estimated $70 million in payments for a mission that is not scheduled to launch until 2016.[23] During the course of this audit, Orbital successfully completed its Antares maiden test flight, which company officials state reduces the risk that the demonstration flight will reveal major technical issues.

Orbital Successfully Completed its Antares Maiden Flight but Has Yet to Demonstrate a Flight to the ISS

Orbital's Antares-Cygnus rocket system is the most complex and technically challenging rocket system the company has attempted. Antares is both the largest rocket the company has built and its first powered by first-stage liquid-fueled engines.[24] In addition, Orbital officials told us that two-thirds of the company's launch system is composed of significantly modified heritage subsystems or new subsystems, which in their view reduces technical risk. However, we have previously reported that the use of heritage technologies often requires significant modification before they are suitable for integration into new products.

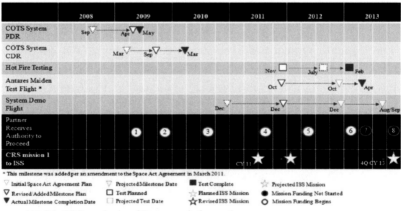

Source: NASA OIG analysis of Program information.

Figure 6. Key Milestones and Test Event Delays for Orbital.

Under the original terms of its Space Act Agreement, Orbital had planned its key milestone – a system demonstration mission to the ISS – for December 2010. Once it became clear that the company would not meet this timeline, NASA amended the Agreement in March 2011 to add a milestone for a maiden test flight of the Antares rocket (see Figure 6). Orbital successfully completed this test flight on April 21, 2013.

Due to technical issues, Orbital's system demonstration flight has been delayed repeatedly and is now scheduled for August or September 2013.[25] Orbital's key milestone delays are as follows:

- *System Preliminary Design Review (PDR)* was delayed 8 months. The PDR is an important review to demonstrate that a system's preliminary design meets all requirements with acceptable risk and within cost and schedule constraints. The PDR establishes the basis for proceeding with detailed design and demonstrates that the correct design option was selected, interfaces have been identified, and verification methods have been described. In its Space Act Agreement, Orbital agreed to develop an unpressurized cargo module; however, NASA's CRS contract calls for pressurized cargo missions. At Orbital's request, NASA modified the Space Act Agreement to include demonstration of a pressurized cargo capability. This major change caused the delay in completing the PDR.
- *System Critical Design Review (CDR)* was delayed 12 months. The purpose of CDR is to demonstrate that the design is sufficiently mature to proceed with full-scale fabrication, assembly, integration, and testing. The CDR should indicate that the technical effort is on track to meet mission performance requirements within the identified cost and schedule constraints. A portion of the delay in getting to CDR was caused by the delayed PDR. The rest of the delay was due to technical issues associated with hardware and electronic systems for the spacecraft. Contrary to industry best practices, Orbital began ordering parts for fabrication of its first two launch vehicles before CDR was successfully completed. Spacecraft designs are not frozen until CDR and are subject to change until that time. Once CDR is completed, all parts specified and designed are ordered and fabricated. Due to early fabrication of the first two launch vehicles and subsequent requirements changes, Orbital's first enhanced cargo module will not be available until Mission 4.[26]

- The *Hot Fire Test* was delayed 15 months. This test was performed on February 22, 2013 to demonstrate the readiness of the rocket's first stage and launch pad fueling systems. The test involved firing Antares' dual AJ26 rocket engines – generating a combined total thrust of 680,000 pounds for approximately 29 seconds – while the first stage was held down on the pad. COTS Program officials told us this delay in testing stemmed primarily from an engine failure during prior testing due to a fuel pipe structural failure caused by stress corrosion, and improvements to the liquid-fuel-capable launch facility at Wallops that were not completed until September 2012.[27]

- The *Maiden Test Flight* of Antares was delayed 18 months. This test flight took place on April 21, 2013, and included a launch of the Antares vehicle carrying a payload simulator in place of the actual Cygnus capsule. The delay for this test flight stems from the Wallops launch pad construction and engine issues mentioned above. After the successful hot fire test, this maiden test flight was Orbital's second-to-last major milestone under its COTS Agreement. In December 2012, NASA granted Orbital authority to proceed with its sixth launch vehicle with payment contingent on completion of this maiden test flight. Despite the delays, the maiden flight of Antares successfully demonstrated all operational aspects of the new launch site and the new rocket, including the ascent to space and delivery of the Cygnus simulator to a target orbit. According to Orbital officials, this reduced significant technical risk associated with the System Demonstration Flight.

- The *System Demonstration Flight* has been delayed at least 33 months. Under Orbital's current plan, the demonstration flight is expected to include: (1) launch of the Cygnus capsule; (2) check-out of Cygnus systems in low Earth orbit; (3) rendezvous, proximity operations, capture, and berth with the ISS; (4) crew entrance and check-out of Cygnus systems while berthed to the ISS; (5) release and departure of Cygnus; and (6) deorbit and re-entry. A large portion of the delay (12 months) can be attributed to the PDR and CDR delays; however, COTS Program officials told us the remaining 21-month delay stemmed primarily from the engine and launch facility issues discussed above combined with systems integration testing that identified issues with individual systems within the Cygnus capsule. In addition, Orbital decided to replace one of the engines on the demonstration launch vehicle due to an internal propellant seal leak observed in prior acceptance testing. Assuming that this flight occurs as planned in August or September 2013, NASA will have paid towards the production of at least six of the eight Orbital vehicles specified in the CRS contract before the company has demonstrated its full flight system.

Under NASA's current schedule, Orbital completed its Antares maiden flight test and plans to undertake its demonstration flight and first cargo mission to the ISS within 7 months. In contrast, SpaceX conducted its Falcon 9 maiden test flight more than 6 months before its first demonstration flight, and an additional 17 months passed between the company's first and second demonstration flights. We acknowledge that technical issues and schedule slippage are common in complex development programs; however, the 3 months between the launch system's August/September demonstration flight and Orbital's first resupply mission to the ISS planned for the last quarter of calendar year 2013 leaves little time to correct any significant issues discovered during the demonstration. Even so, Orbital officials stated that a 3-month turnaround time is achievable because the company has considerable experience in launching multiple missions in a short timeframe and have the space systems readily available.

Continued Funding of Orbital's CRS Contract Despite Substantial Launch Delays Increases NASA's Financial Risk

NASA increased its financial risk by funding production of multiple Orbital spaceflight systems while Orbital's development efforts were still under way. According to best practices for space system acquisition, production of a flight system should not occur until the system has been successfully demonstrated in a relevant environment.[28] The purpose of these practices is to minimize the concurrency between product development and production of a system.[29] Previous reviews by the NASA OIG and the Government Accountability Office (GAO) have found that committing to production before system development is complete is a high-risk strategy that often results in performance shortfalls, cost increases, schedule delays, and test problems.[30]

We understand that NASA accepted the risks of concurrent development with both SpaceX and Orbital out of a need to ensure a redundant cargo capacity to meet the ISS resupply schedule, and we do not second guess the Agency's decision to concurrently fund up to three spaceflight systems for each company. However, in the case of Orbital, we believe that NASA has leaned too far forward by continuing to fund the company's CRS missions when Orbital did not meet major developmental milestones specified in its Space Act Agreement, which in turn delayed launch dates for its CRS resupply missions.

In the Orbital CRS contract, each cargo mission has 10 milestones, the first 7 of which are tied to rocket system fabrication for items such as long lead procurement and integration and testing. The final three milestones, representing 30 percent of the remaining funds, are dependent on successful launch activities such as cargo integration review, launch, and delivery. As a result, NASA is paying Orbital 70 percent of the funds associated with the company's first three CRS missions and 40 to 50 percent of the funds associated with Missions 4 and 5 – all before Orbital demonstrates that its system can successfully launch and rendezvous with the ISS.

As of the end of FY 2012, NASA paid Orbital $633 million under the CRS contract toward activities associated with building five rocket systems, three of which were substantially complete by December 2012.[31] This funding represents 33 percent of the total funding NASA agreed to pay Orbital for its 8 resupply missions. In contrast, NASA paid SpaceX approximately $100 million toward three missions under its CRS contract prior to the company's Falcon 9 maiden test flight, and a total of about $300 million prior to its successful demonstration flights. Table 3 summarizes the funding Orbital has received from NASA.[32]

Table 3. Summary of COTS and CRS Funding for Orbital

	Agreement type	Total award (millions)	Total spent [a] (millions)	Percent spent
Commercial Orbital Transportation Services (COTS)	Space Act Agreement	$288	$276	96%
Commercial Resupply Services (CRS)	Fixed-price contract	$1,900	$633	33%
	Total	$2,188	$910	42%

[a] Dollars as of the end of FY 2012.
Source: NASA OIG analysis of Program information.

We question NASA's decision to continue funding milestones for up to six Orbital CRS missions given the financial risk to the Agency if major modifications are required to the company's launch system after the demonstration flight. Funding so many missions in advance of a demonstration flight increases NASA's financial and programmatic risk. Under the original

CRS contract, NASA was scheduled to fund work on three missions before Orbital completed its demonstration flight. However, even though most key development milestones under the Space Act Agreement and therefore the launch dates for Orbital's CRS flights have been delayed, NASA only recently adjusted Orbital's CRS payment schedule to reflect these delays.[33]

Moreover, as a result of the schedule delays and its decision not to adjust the payment schedule accordingly, NASA has funded all of Orbital's missions months or even years earlier than necessary to fly to the ISS, according to the Agency's revised flight schedule.

Standard lead times for development of similar space systems is 24-30 months, and the SpaceX and CRS contracts specify lead times consistent with this standard. However, NASA funded and Orbital began fabricating systems for the company's first six missions an average of 41 months before they are now scheduled to fly. Figure 7 summarizes the relationship between fabrication and current mission schedules.

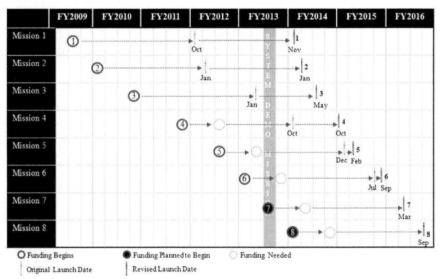

Source: NASA OIG analysis of Program information.

Figure 7. CRS Funding and Fabrication Schedule.

In our judgment, funding Missions 1, 2, and 3 was reasonable despite the program delays and schedule changes because NASA officials had a need to start CRS missions to the ISS immediately following COTS demonstration flights. However, we question NASA's decision to pay Orbital approximately

$150 million in FY 2011 through FY 2013 toward the systems for Missions 4 and 5, which are not scheduled to fly until FY 2015. At the time these funding decisions were made, NASA Program officials were well aware of the extent of slippage in both Orbital's COTS milestones and its CRS launch dates.

Furthermore, so far in FY 2013, NASA has spent approximately $10 million on Mission 4, and we estimate the Agency will spend an additional estimated $70 million on Missions 4, 5, and 6 by the end of FY 2013. In addition, Orbital has requested as part of their Launch on Need proposal to begin work on Mission 7 in May 2013 and Mission 8 in November 2013. We estimate that should NASA ultimately agree to this request it would constitute a premature expenditure of an additional estimated $70 million in FY 2013 and roughly $60 million in FY 2014, based on NASA's current ISS flight manifest.

Table 4 summarizes the amounts NASA has paid or is scheduled to pay Orbital per mission under the CRS contract. In our judgment, delaying authority to proceed until closer to when missions are scheduled to launch would better protect NASA's financial interest against the possibility that Orbital is ultimately unable to successfully demonstrate its launch system or the demonstration flight shows that major modifications to the system are required.

Table 4. Summary of Potential CRS Savings per Fiscal Year

	FY 2011 (millions)	FY 2012 (millions)	FY 2013 (millions)	Additional Request for FY 2013 (millions)	FY 2014 (millions)	Total (millions)
Spent Dollars That Could Have Been Deferred	$45	$95	$10			$150
Dollars to Defer from FY 2013 to FY 2014			$70	$70		$140
Dollars to Defer from FY 2014 to FY 2015					$100	$100

Source: NASA OIG analysis of Program information.

Balancing and Managing Financial Risk

NASA officials acknowledged the financial risk associated with funding Orbital's development of spaceflight capabilities while concurrently funding the fabrication of six launch vehicles. However, in their view the

programmatic risks of not having reliable transportation capabilities to service
the ISS outweighed the financial risk. To this end, NASA program officials
said they funded Orbital for six missions because NASA required a Launch on
Need capability in case cargo resupply missions from either SpaceX or the
Agency's international partners became unavailable. In addition, NASA
Program officials said upgrades to Orbital hardware on Missions 4 through 8
required a longer lead time to build.

NASA officials also explained their rationale for accepting this level of
financial risk as a way to ensure a second private company was financially
"healthy" enough to participate in the CRS activities. The officials said that
despite the milestone delays under Orbital's Space Act Agreement, they
currently see no major technical problems with the company's launch system
or capsule. Further, NASA Program officials noted that to reduce risk, they
have linked 20 percent of the payment for each mission to milestones Orbital
can meet only by successfully completing deliveries to the ISS. In addition,
they said they have been holding both SpaceX and Orbital to their original
mission pricing as a means of holding the contractors accountable for the
schedule delays, which helps reduce the Government's financial risk.

According to Orbital officials, the successful Antares maiden flight in
April 2013 has reduced the risk that the full demonstration flight will reveal
major technical issues with the company's system. Given this flight and the
completion of ISS integration, which took place in March 2013, Orbital
officials said they have demonstrated the capability to execute ISS resupply
missions. In addition, Orbital officials stress that because the CRS resupply
contract is fixed-price, any technical changes that result in additional costs will
be borne by Orbital alone. They also point out that in the event the contract
needs to be terminated for cause, all CRS payments are recoverable.

Nevertheless, we maintain that by buying services – valued at almost $1
billion – for a system that has not been fully demonstrated, NASA has incurred
an unnecessary risk. In our view, continuing work on Orbital's fourth and fifth
rocket systems and beginning work on Missions 6 through 8 in the absence of
a successful system demonstration flight introduces unnecessary financial risk
to NASA, particularly given that the start-work dates for these missions are
well in advance of the current launch schedule. The current manifest indicates
that Orbital's Mission 6 is not scheduled to launch until FY 2015, with
Missions 7 and 8 not scheduled to launch until at least FY 2016.

Despite their confidence in Orbital's system, Program officials
acknowledged our concern about the level of financial risk NASA was
accepting given that the company's system demonstration flight has slipped to

August or September 2013. In addition, during the course of our audit we discussed with NASA and Orbital officials our concerns regarding the Agency's increased financial risk associated with paying towards rocket systems so far in advance of when they are needed to meet the ISS flight manifest. NASA officials generally agreed with our assessment, and took action to reduce the Agency's financial risk. For example, NASA officials enforced a section of the CRS contract that enables them to hold funding for Missions 4 and 5 at 50 percent as a result of launch schedule delays. Furthermore, although NASA granted Orbital authority to proceed with Mission 6 in December 2012, payments for the mission were withheld until the completion of the Antares maiden test flight.

While we acknowledge these positive steps toward mitigating NASA's financial risk, we believe that going forward, NASA should ensure that contractual plans and agreements are updated to reflect the lead times required to meet revised launch dates. If launch dates slip, NASA should adjust the contracts to ensure that the authorized lead times – and NASA payments – reflect the revised schedules.

Conclusion

Since 2006, NASA has worked with its commercial partners to develop commercial capabilities to transport cargo to low Earth orbit. In 2008, NASA entered into contracts with two companies to utilize those capabilities by delivering cargo to the ISS. To date, SpaceX has successfully completed the COTS Program and flown two CRS missions to the ISS, while Orbital prepares to demonstrate its complete flight system. While we are encouraged by the successful maiden test flight of Orbital's Antares rocket, we remain concerned about the financial risk NASA is taking by funding systems for Orbital missions so far in advance of expected launch dates and before the company has completed a total spacecraft system demonstration flight to the ISS. With six systems either substantially complete or in production, NASA has invested considerable Agency resources in Orbital's current system design prior to a successful demonstration flight.

Consequently, we question NASA's decision to continue to fund or begin funding Orbital's fourth, fifth, and sixth resupply missions. While we acknowledge that Orbital's testing to date is meeting expectations, if the system ultimately is unable to successfully launch, rendezvous, and berth with the ISS, NASA may not be able to recover the approximately $150 million it has invested in these later missions.[34] In addition, if NASA executes its current FY 2013-2014 spending plan, it will prematurely authorize payment of an additional estimated $240 million to Orbital for these missions. We believe that this level of risk is unnecessary given that NASA has other options for resupplying the ISS, including SpaceX and other international partners.

Recommendations, Management's Response, and Evaluation of Management's Response

We recommended that going forward, the NASA Associate Administrator for the Human Exploration and Operations Mission Directorate ensure that contractual agreements for the commercial cargo providers are updated as appropriate to reflect lead times required to meet new launch dates. If launch dates slip, NASA should adjust contract work plans to ensure that the authorized lead times – and NASA payments – reflect the revised schedules.

Management's Response. Although the Associate Administrator disagrees that NASA has accepted too much financial risk in the way it has implemented the Orbital CRS contract, he nevertheless concurred with our recommendation to ensure CRS contracts are updated to reflect the lead times required to meet any revised launch dates. He stated that NASA will work closely with CRS contractors to update work plans in an iterative process, including reviewing and updating the plans at quarterly meetings, and will document the logic behind any decision not to update a particular work plan. He noted that the next quarterly meetings are scheduled for July for SpaceX and August for Orbital.

With regard to our finding regarding financial risk, the Associate Administrator stated that NASA determined that the programmatic risks of not starting hardware development needed for cargo resupply were substantially greater than the financial risks posed to the Agency by doing so. He further stated that NASA uses existing payment cap protections and other contractual provisions to reduce financial risks and align payments with technical performance.

Evaluation of Management's Response. Management's proposed actions are responsive; therefore, the recommendation is resolved and will be closed upon completion and verification of the corrective actions.

With regard to the Associate Administrator's comments regarding our finding of financial risk, we agree that balancing programmatic and financial risk is critical to ensure the success of the commercial cargo program. However, as outlined in the report, we continue to believe that NASA has been too slow to adjust its payment schedule to Orbital given the substantial slippage in the launch schedule for the company's resupply missions.

Appendix A

Scope and Methodology

We performed this audit from July 2012 through April 2013 in accordance with generally accepted government auditing standards. Those standards require that we plan and perform the audit to obtain sufficient, appropriate evidence to provide a reasonable basis for our findings and conclusions based on our audit objectives. We assessed that the evidence obtained provides a reasonable basis for our findings and conclusions based on our audit objectives. This audit focused on the COTS Program managed by NASA's Commercial Spaceflight Development Office, and the CRS contract managed by NASA's International Space Station Program, both within the Human Exploration and Operations Mission Directorate. To determine the Agency's management of the COTS Program and CRS contracts, progress made, and challenges hindering the successful implementation of the Program and contracts, we reviewed laws, regulations, and policies in order to determine compliance with required guidance and best practices. We obtained and reviewed prior reports related to NASA's ability to address the development and collaboration challenges of the COTS Program. We interviewed key personnel within NASA's Commercial Spaceflight Development Office and the COTS Program located at NASA Headquarters, Kennedy Space Center, Johnson Space Center, and commercial partners at their corporate sites. To determine whether both the commercial partners were on track to provide a system capable of resupplying the ISS, we analyzed schedule timelines and partner performance under both COTS and CRS to date. We reviewed CRS contracts and interviewed key NASA personnel to determine whether NASA was properly managing the work plans for CRS. We compared contract payment milestones to milestones in the Space Act Agreements and in NASA guidance and analyzed the differences. To determine whether the Space Act Agreements used in cargo development are readily adaptable to other programs such as commercial crew development, we interviewed program managers and officials from both the commercial crew and cargo programs, along with their supervisors and supporting teams.

Use of Computer-Processed Data. We used computer-processed data to perform this audit. We collected computer-processed milestone payment cost data for the COTS Program and the CRS contracts from the beginning of the Program through the end of FY 2012. Program officials downloaded the data from NASA's financial management program and provided the data in

Microsoft Excel. In order to verify the accuracy of this data, we corroborated the information provided with documentation such as the original COTS Space Act Agreements and amendments, the CRS contracts, task orders, work plans, and budget data. We analyzed this data to enable us to evaluate partner and contractor performance, identify risks to the Program, and quantify NASA's investment to date. We assessed that the cost data we received was sufficiently reliable, but we did not rely solely on the computer-processed data to support our findings, conclusions, or recommendations.

Review of Internal Controls

We reviewed NASA policies and procedures to determine internal control requirements for management of the COTS Program. The results of this review will be provided in a separate memorandum to NASA management.

Prior Coverage

During the last 5 years, the NASA OIG and the GAO have issued 11 reports or testimony of particular relevance to the subject of this report. Unrestricted reports can be accessed over the Internet at http://oig.nasa.gov/audits/reports/FY13 and http://www.gao.gov.

NASA Office of Inspector General
"NASA's Challenges to Meeting Cost, Schedule, and Performance Goals" (IG-12-021, September 27, 2012)
"NASA's Challenges Certifying and Acquiring Commercial Crew Transportation Services" (IG-11-022, June 30, 2011)
"Review of NASA's Acquisition of Commercial Launch Services" (IG-11-012, February 17, 2011)

Government Accountability Office
"Commercial Space Launches: FAA Should Update How It Assesses Federal Liability Risk" (GAO-12-899, July 30, 2012)
"Commercial Space Transportation: Industry Trends, Government Challenges, and International Competitiveness Issues" (GAO-12-836T, June 20, 2012)

"Missile Defense: Opportunity Exists to Strengthen Acquisitions by Reducing Concurrency" (GAO-12-486, April 2012)

"NASA: Significant Challenges Remain for Access, Use, and Sustainment of the International Space Station" (GAO-12-587T, March 28, 2012)

"NASA: Assessment of Selected Large Scale Projects" (GAO-12-207SP, March 1, 2012)

"National Aeronautics and Space Administration: Acquisition Approach for Commercial Crew Transportation Includes Good Practices, but Faces Significant Challenges" (GAO-12-282, December 15, 2011)

"Key Controls NASA Employs to Guide Use and Management of Funded Space Act Agreements are Generally Sufficient, but Some Could Be Strengthened and Clarified" (GAO-12-230R, November 17, 2011)

"NASA: Commercial Partners Are Making Progress, but Face Aggressive Schedules to Demonstrate Critical Space Station Cargo Transport Capabilities" (GAO-09-618, June 16, 2009)

APPENDIX B

Commercial Partner Table

Table 5. List of COTS Recipients and CRS Awards

Company	Orbital Sciences Corporation (Orbital)	Space Exploration Technologies Corporation (SpaceX)
System	Cygnus/Antares	Dragon/Falcon 9
Total COTS Funds Available	$288 million	$396 million
Total COTS Funds Used (Percentage of Total Funds)	$276 million (96%)	$396 million (100%)
Total Potential Value of CRS Contract	$1.9 billion	$1.6 billion
Total CRS Funds Used (Percentage of Total Funds)	$633 million (33%)	$462 million (29%)
Capability	Internal pressurized cargo delivery and disposal	Internal pressurized cargo delivery return; unpressurized cargo delivery and disposal
Payload to ISS	1,700 kg (standard) 2,700 kg (enhanced)	3,310 kg
Payload from ISS	2,000 kg (disposed)	3,310 kg (returned, pressurized + unpressurized)
Number of Flights Ordered	8	12
Number of Flights Completed	0	2
Planned Missions by Fiscal Year	2013 – 2 2014 – 2 2015 – 2 2016 – 2	2012 – 1 2013 – 2 2014 – 3 2015 – 3 2016 – 3

Source: NASA OIG analysis of Program information.

Final report released by:
Paul K. Martin Inspector General

Major Contributors to the Report:
Ridge Bowman, Director, Space Operations Directorate
Kevin Fagedes, Project Manager
Letisha Antone, Auditor
Cedric Campbell, Associate Counsel to the Inspector General –
Central Region
Michael Palinkas, Management Analyst
Janice Smith, Auditor
Troy Zigler, Program Analyst

End Notes

[1] By law, the ISS must be maintained until at least 2020; however, NASA is examining and Congress is considering the feasibility of extending the Station's life.

[2] Space Act Agreements are a form of "Other Transaction Authority" provided in the National Aeronautics and Space Act of 1958 (as amended) that allows NASA to establish a set of legally enforceable commitments between the Agency and a partner per NASA Policy Directive 1050.1I, "Authority to Enter Into Space Act Agreements," December 23, 2008.

[3] SpaceX's transportation system includes a rocket booster system (Falcon 9) and capsule (Dragon), while Orbital's system includes a rocket booster (Antares) and capsule (Cygnus).

[4] NASA has issued task orders and work plans under the FAR-based contracts that outline schedules, payment information, and milestones for the 20 resupply missions. To date, NASA has given SpaceX authority and partial funding to proceed with 7 missions and Orbital authority and partial funding for 6 missions.

[5] The Office of Inspector General is also conducting an audit of NASA's Commercial Crew Program and will issue a report later this year.

[6] As of May 2013, Orbital officials stated they had received 70 percent of payments for Missions 1 through 3, 50 percent for Mission 4, and 40 percent for Mission 5.

[7] As a point of comparison, NASA had funded or partially funded four missions before SpaceX's first demonstration mission and five missions before the second demonstration mission.

[8] As of June 2013, NASA officials informed us that they have delayed authority to proceed for Mission 7 due to slippages in the CRS launch schedule.

[9] Per the CRS contract, ISS integration is "the activities required to ensure that SSP 50808 (ISS requirements document) have been met; necessary hardware and software developments to interface with the ISS have been completed; and joint on-orbit integrated operations plans have been finalized." Orbital reported completing ISS integration in March 2013.

[10] As defined in the Federal Acquisition Regulation, a task order is an order for services defined in a contract. NASA uses a document called a work plan to describe the ways in which the contractor will execute the task order.

[11] We use the term "partner" to describe the commercial companies that perform work with NASA under a Space Act Agreement.

[12] NASA Policy Directive 1050.1I, "Authority to Enter into Space Act Agreements," December 23, 2008.

[13] SpaceX's Dragon capsule returns to Earth, while Orbital's Cygnus capsule will burn up upon reentry and therefore can only be used to dispose of unneeded materials from the ISS.

[14] Examples of task orders include conducting studies and designing new equipment for future missions.

[15] In accordance with its Space Act Agreement, Orbital began building an unpressurized capsule in February 2008. In December 2008, Orbital accepted a contract award for the CRS missions with a requirement for a pressurized capsule, which necessitated a change in design and delayed development of its system. This issue is discussed in more detail in the Results section.

[16] The Russian Space Agency (Roscosmos) uses its "Progress" vehicle to resupply its segment of the ISS, but the United States does not have a current resupply agreement with Russia.

[17] Barter agreements involve the exchange of goods or services rather than money. The framework for cooperation was established in the January 1998 ISS Intergovernmental Agreement with specific details for barter agreements in subsequent Memorandums of Understanding between NASA and partner countries.

[18] Under the original Space Act Agreement, the final SpaceX COTS demonstration was the only flight that included an actual berthing with the ISS.

[19] While the Dragon can carry 3,310 kilograms of pressurized up mass, the amount of mass carried is dependent on the cargo's density given the available volume. Furthermore, previous resupply missions by international partners had reduced the amount of cargo that SpaceX needed to deliver to the ISS on this first mission. The higher priority need for this mission was to return cargo back to Earth.

[20] A planned second burn of the Falcon 9 second stage was cancelled by NASA in order to meet ISS safety requirements. This burn would have been needed to place the satellite into the correct orbit. While the second stage had sufficient propellant for the second burn, the amount available was just below the level NASA required to meet its acceptable risk criteria.

[21] Hardening against the effects of radiation involves using special radiation-resistant materials when fabricating computer processors and components, and then insulating and shielding the systems from the effects of radiation.

[22] Orbital officials stated that as of May 2013 the company has received 70 percent of the payments associated with Missions 1 through 3, 50 percent for Mission 4, and 40 percent for Mission 5.

[23] Launch on Need capability means that the contractor has a "rolling spare" vehicle readily available in the event NASA needs a launch earlier than is planned in the current launch schedule. As of June 2013, NASA officials informed us that they have delayed the Authority to Proceed for Mission 7 due to slippages in the CRS launch schedule.

[24] There are two basic types of rockets: solid propellant fueled and liquid propellant fueled. Liquid-fueled rockets are considered more complicated to operate because they require pumps and fuel lines, which tend to make them heavier than solid-fueled engines. Although a liquid-fueled rocket has its complications, there are also advantages. For example, unlike a solid-fueled rocket, which, once ignited, burns until all fuel has been exhausted, a liquid-fueled rocket has valves and pipes that direct (or shut off) the fuel, making it easier to control the amount of the thrust.

[25] Orbital originally proposed in its COTS agreement to develop, test, and demonstrate the entire system in 2 years. NASA Program officials told us they knew this was an optimistic and aggressive schedule and would likely slip, but they nonetheless approved Orbital to work toward that goal.

[26] The enhanced cargo module provides the capability to increase delivery of cargo to the ISS from 1,700 kilograms to 2,700 kilograms. NASA and Orbital originally planned to use an enhanced cargo module for Mission 3; however, subsequent changes in requirements associated with the vehicle visiting the ISS made it necessary to slip the enhanced capability to Mission 4.

[27] The launch facility at Wallops, known as the Mid-Atlantic Regional Spaceport, is overseen by the Virginia Commercial Space Flight Authority.

[28] GAO, "Best Practices: Better Management of Technology Development Can Improve Weapon System Outcomes" (GAO/NSIAD-99-162, July 30, 1999). Technology readiness level 6 – demonstration of a prototype in a relevant environment – is the level of technology maturity that constitutes low risk for beginning system fabrication for production models.

[29] Concurrency is broadly defined as the overlap between technology development and product development or between product development and production. While some concurrency is understandable, committing to product development before requirements are understood and technologies mature or committing to production and fielding before development is complete is a high-risk strategy that often results in performance shortfalls, unexpected cost increases, schedule delays, and test problems. It can also create pressure to keep producing to avoid work stoppages. GAO, "Missile Defense: Opportunity Exists to Strengthen Acquisitions by Reducing Concurrency" (GAO-12-486, April 2012).

[30] GAO, "NASA: Assessments of Selected Large-Scale Projects" (GAO-12-207SP, March 1, 2012) and NASA OIG, "NASA's Challenges to Meeting Cost, Schedule, and Performance Goals" (IG-12-021, September 27, 2012)

[31] The payments to Orbital are used for all costs related to the entire program including spacecraft hardware, ground systems, and program reviews.

[32] As a point of comparison, SpaceX began work on flight hardware for its first three missions before its Falcon 9 maiden test flight in June 2010. At the same point in its development cycle – prior to its April 2013 Antares test flight – Orbital had begun work on six of its eight missions.

[33] Although there are no contractual ties between the CRS contract and the COTS Space Act Agreements, in December 2012 NASA negotiated with Orbital to tie payment for Mission 6 to the completion of the Antares maiden test flight.

[34] NASA officials have stated that CRS payments are commercial interim payments that would be fully recoverable in the event of termination of the entire contract for cause (per FAR 52.212-4[m] "Termination for Cause"). However, our analysis indicates that Orbital will receive additional funds associated with the contract before the demonstration flight now scheduled for August or September 2013. It may be very difficult to recover these funds in the event of a Termination for Cause.

INDEX